COWBOY CAVE

NUMBER 104 1980

COWBOY CAVE

by

Jesse D. Jennings

Sections by

R.N. HOLMER, D.G. WEDER, N.J. HEWITT, J.C. JANETSKI,
W.A. LUCIUS, F.W. HULL, J.E. DODGE,
F.W. HULL AND N.M. WHITE,
AND P.R. BARNETT AND N.J. COULAM

Appendices by

W.G. SPAULDING AND T.R. VAN DEVENDER, W.G. SPAULDING AND K.L. PETERSEN,
R.M. HANSEN, J.C. WINTER, B.J. ALBEE, P.F. HOGAN, AND L.W. LINDSAY

UNIVERSITY OF UTAH
ANTHROPOLOGICAL PAPERS

Jesse D. Jennings, Editor
Sharon S. Arnold, Associate Editor

University of Utah Press
Salt Lake City, Utah

CONTENTS

ILLUSTRATIONS

TABLES

ACKNOWLEDGMENTS

The research reported here was made possible with the help of the Bureau of Land Management in the persons of Richard Fike, BLM State Archeologist, and David Hunsaker, BLM Recreation Specialist from the Price, Utah, Office, who assisted in issuing the permit. Both expressed their interest in the project by visiting the site during the progress of the excavation. Equally helpful in a score of ways was James Walters, then District Ranger of Canyonlands National Park in charge of the Maze District of the Glen Canyon Recreation Area and the Barrier Canyon portion (detached) of Canyonlands National Park. Through his generosity a campsite, emergency radio contact facilities, the District water supply at French Springs, mail service, and District shop facilities were made available for the field school operation. His day-to-day interest included frequent trips to the site during excavation. Another Canyonlands National Park employee whose help contributed to a successful season was Sandra Jarrett, who accepted and relayed via Canyonlands National Park radio several emergency messages required during the early days of camp operations.

Student participants from the University of Utah were: *graduates*, Jeffery Colville, Alan Lichty, Leonard Losee, William Lucius; *undergraduates*, Nancy Coulam, William Fushimi, Peggy Barnett, Kendel Malstrom, and Mark Marzolf. From out-of-state schools there were: James Dodge, University of Oregon; Adrien Hannus, South Dakota State University; Susan Pollock, Cornell University; Nancy White, Albion College, Michigan.

Most of the Utah students later participated actively in the analyses of data. Their work, in general, was well done. Their contributions are acknowledged in the bylines to various sections that follow. Robert Stuckenrath performed radiocarbon assays and offered many helpful comments toward their interpretation. Also contributing to analyses were Jane Jennings, weaver and textile specialist, who participated for many weeks in the analysis of all the fiber artifacts (sandals, basketry, cordage, and knots) and made several sketches used as illustrations; Dr. John Moore, who tested for human blood on the fiber pads; Carol Weins, who identified the woody plant specimens; Beverly Albee, who identified plant fibers as well as many seeds and verified the seed identifications cited in Table 2 of Appendix II; William Behle and John Wycoff, who identified the feathers; Wade Miller, who identified the bones of extinct animals; Tommy D. Moore, who identified bison hair; Laurence Kittleman, who identified the sandstones utilized in artifacts; and Beverly and Howard Albee, who prepared the floral inventories (*Appendices VII* and *VIII*). David Crompton and Richard Holmer conducted an

intensive six-day tutorial session in
photography and topography during the
first week of the session.

Acknowledgments to other experts who
assisted with analyses of data reported
appear in the proper context in the
several appendices.

Jesse D. Jennings

April 1977

COWBOY CAVE

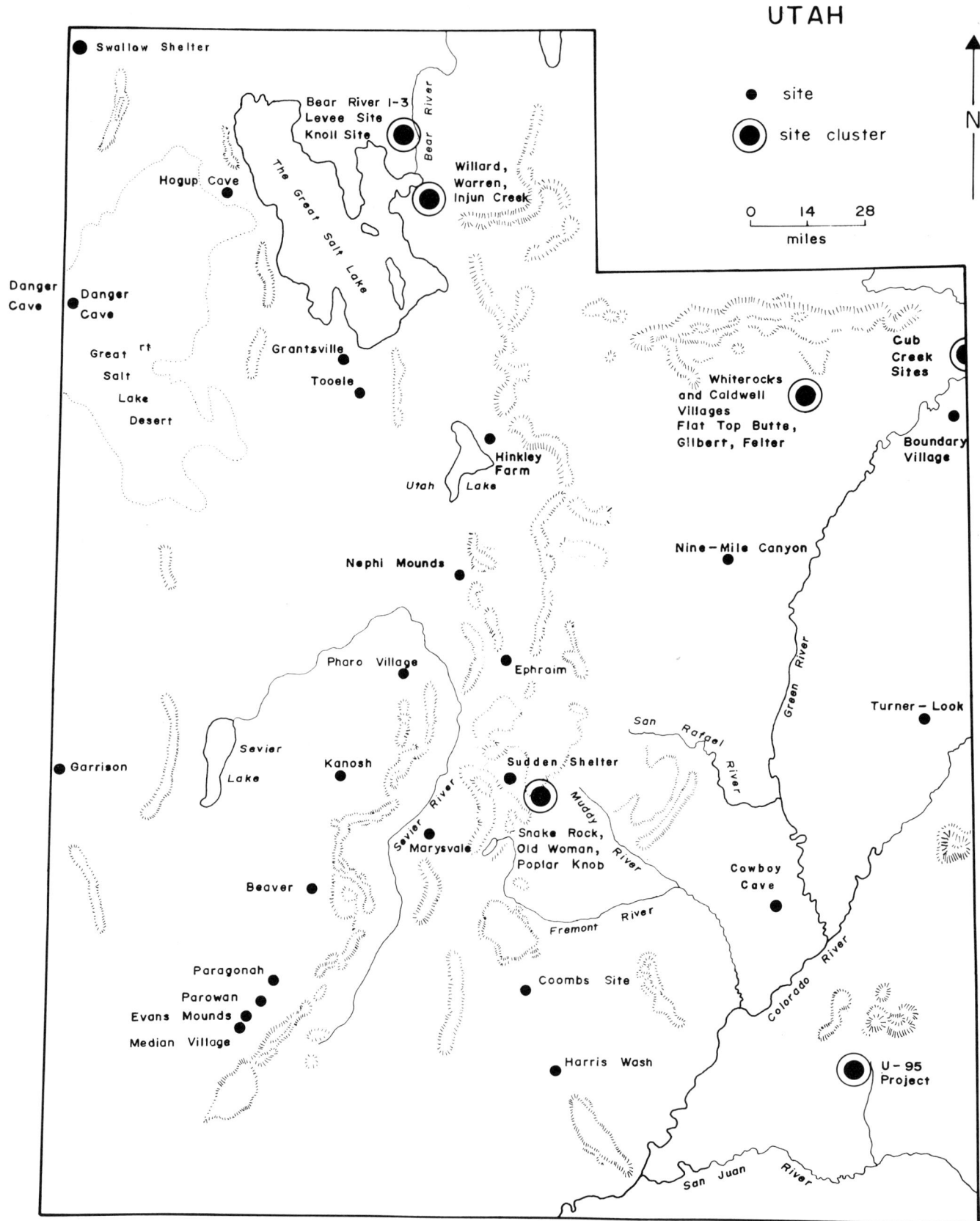

FIG. 1. Locations of archeological sites in Utah.

INTRODUCTION

Cowboy Caves (42Wn420 and 42Wn421), now designated Cowboy Cave (42Wn420) and Jim Walters Cave (42Wn421), were brought to the attention of the University of Utah Archeological Survey in the spring of 1973 by Edward McTaggart of the Price, Utah, Office of the Bureau of Land Management as a matter of routine. Inspection of the site was followed by a test cut. The site proved to contain a rather thick deposit of cultural debris as was originally reported as being probable. Moreover, the test cut revealed that the cultural deposit sealed a thick stratum of "chopped" vegetation identifiable as herbivore dung. The initial identification as sloth dung proved to be incomplete; there was also dung from bison, elephant, horse, and a camel-like ruminant (see *Appendix III*).

The importance of a site where association of man with extinct megafauna might reasonably be expected, led to its selection as the object of study for the annual field school for training in archeological field procedures and techniques of the University of Utah Department of Anthropology. The 1975 summer session was held there. Accordingly, thirteen students, most from the University of Utah, began work there on June 6, 1975. The school was, as usual, directed by Jesse D. Jennings, who was assisted that season by Alan Schroedl. More than 500 man-days were devoted to

the excavation of Cowboy Cave, with some 30 to 40 man-days going toward a testing of Walters Cave. When it was determined that the caves appeared to be nearly identical in stratigraphy and content, work in Walters Cave was terminated, allowing all personnel to concentrate on Cowboy. It was therefore possible to remove the full cultural deposit in the latter cave; Walters Cave had been saved, it was thought, for any desirable or necessary future study. Unfortunately, this smaller cave was later almost totally destroyed by an organized band of vandals, as yet unidentified. In Cowboy Cave, more than 50 percent of the dung deposit, probably in excess of 10 cu. m., remains for later study. The dung has interested several scholars in other fields. For example, Richard Hansen, range management specialist, and Paul S. Martin, a paleontologist/palynologist who is also deeply involved with attempting to understand the complex factors leading to the extinction of many megafaunal species, have evinced great interest.

Excavation was conducted under Department of the Interior Permit No. 74-UT-011, issued by the Bureau of Land Management. Being on BLM lands, the caves were believed to enjoy a certain measure of protection from vandals. (Experience proved this expectation to be in error.) The sites have been nominated for the National Register of Historic Places.

LOCATION

Cowboy Cave is in Wayne County, Utah, on the western edge of the Canyonlands province of the Colorado Plateau. It is on the north bank of a short, waterless, unnamed tributary of Spur Fork canyon. Spur Fork trends north to join Barrier Creek in Horse Shoe canyon. The geographic location of the caves is approximately 38° 19' 04" North latitude, 110° 12' 08" West longitude.

SETTING

The situation the two caves occupy exemplifies the spectacular beauty of the canyon country wherever the pink sculptured domes and cliffs of Navajo sandstone and the extensive dunes of sand eroded from the same formation are the dominant landscape features. The twin caverns are on the north bank of the short canyon, a few meters above the canyon floor. They open to the southeast, are therefore well lighted in summer, and receive considerable direct sunlight during afternoons in the winter months. Less than 50 m. downslope south from the caves, there is a weak seep that is evidently the only permanent water in the canyon. The seep appears to issue from the seam between the Navajo formation and the underlying Kayenta. It is reported never to dry up entirely. Vegetation in the lowest layers testifies to perennial stream flow as late as 7000 B.P. Southeast about 100 m. there is another water source, a very large, deep natural reservoir or pothole which contains several hundred gallons of water in the spring and after every rain.

Although no specific records for the cave area exist, the climate can only be described as one of rigorous desert nature, judging by the Canyonlands National Park seven-year record from Island in the Sky District, only 36.5 km. away. The elevation of Island in the Sky is 6,200 ft. (1,890 m.), slightly higher than Cowboy Cave at 5,800 ft. (1,770 m.). The record shows an annual mean precipitation of 20± cm. (8.0± in.), with about half that amount coming between May 1st and September 30th. Average temperatures are lowest in January at -7° C; highest in July at 33° C. Average extreme ranges show -16° C for January, and 37° C in July (see Table 1).

The caves are carved into a low, Navajo sandstone cliff. Both are more or less symmetrical grottoes with gracefully arched, low ceilings and smooth walls. The sandstone floors of both caves rise toward the backs of the caves, the difference in elevation between the portal and the rear being 1.5 m. The floor is a thin, creamy-white layer occasionally noted in the lower measures of the Navajo sandstone. In this case it appears to be markedly more resistant, even projecting slightly beyond the portals to create a shelf or lip overhanging the layers beneath which have eroded slightly from under it.

Floral resources appear to be about what they have been for at least 11,000 years, with the exception of certain grasses which are mentioned and discussed in *Plant Macrofossil Analysis,* and numerous now-absent conifers listed in *Appendix I.* The plant community can be described in gross terms as the pinyon-juniper complex typical of southeastern Utah at this elevation; the cave floors lie almost exactly 1,700 m. above sea level. The mesas and "flats" adjacent to the canyon therefore comprise the typical upland grass community, dominated here by *Muhlenbergia pungens* (?) and *Ephedra nevadensis.* Scattered small pinyon and juniper dot the sandy, grassy areas. Granting the past and present economic importance of the dominant grasses, there remain a number of other plants in both the canyon community and the upland community useful to an aboriginal population as raw material for manufactures and food resources. The more than 100 species in the canyon include yucca, cliff rose, cottonwood, willow, oak, cactus, many composites, Rocky Mountain bee plant, and barberry, as well as the pinyon and juniper. Many of the same species occur on the uplands. Some edible forbs occur there as well.

FIG. 2. View west from mouth of Cowboy Cave. Seep is located at base of low cliff slightly left of center.

TABLE 1

AVERAGES OF RECORDED TEMPERATURES AND PRECIPITATION, 1968-75
ISLAND IN THE SKY

| | Temperatures | | | | Average Precipitation |
	Average Extreme Maximum	Average Extreme Minimum	Average Maximum	Average Minimum	
January	11° C	−16° C	2° C	−7° C	.74 cm.
February	14	−11	6	−3	.60
March	20	−8	12	0	1.2
April	22	−5	16	3	1.9
May	31	1	24	10	1.3
June	35	4	29	14	1.8
July	37	14	33	18	2.3
August	35	11	31	17	2.2
September	32	4	25	12	1.1
October	26	−4	17	6	4.0
November	15	−7	9	0	1.8
December	10	−13	3	−6	1.3

The full inventories of the two communities are listed in *Appendices VII* and *VIII*.

The locale also provides abundant stone for millstones, these being derived from the tough, thin-bedded, gray purple Kayenta into which the canyon has cut its lower channel--gray lenticular inclusions that were evidently superior millstone material. No ledges, lenses, or outcrops of the glassy minerals used for chipped stone tools--knives, points, scrapers, or drills--were noted in the immediate vicinity. However, to the east, no more than 15 to 20 km. away in the Maze District, there are cobble fields, ledge outcrops, etc., where cherts, varying from brown through bright red to mottled gray, occur (Lucius et al. 1976). The Maze is deemed the probable source of the limited quantity of chipped stone and debitage recovered at Cowboy Cave.

SITE DESCRIPTION

The sites, already identified as twin caverns of identical elevation at 1,700 m., have large symmetrical portals rising about 5 m. above the cave floors. Walters Cave is 15 m. in length or depth, by 11 m. wide at the mouth. Cowboy, much larger, is 33 m. long by 12 m. wide. The floor area of Walters is thus 165± sq. m., with the floor of Cowboy being 400± sq.

FIG. 3. Walters and Cowboy Caves (left of center) from base of cliffs on south side of canyon. Walters Cave is on the left. Overhangs to the right contained no cultural debris.

FIG. 4. Closer view of two caves.

m., about two and a half times larger than Walters (Figs. 5 and 6). When discovered, the mouths of both caves were somewhat blocked by sandstone spalls of varying size and what appeared to be rather deep sand dunes.

The dunes and spall effectively masked the presence of extensive cultural deposits, except that an occasional surface fragment or shred of juniper bark hinted at some degree of human use. The dunes, presumed to be wind deposited, were located east of the central long axis of the caves, with little surface sand west of the median axis. The deep, eastern sands were probably deposited by the prevailing southwesterly wind eddying in the cave mouths. In Cowboy Cave, there was an extensive crater 3 or 4 m. in dia. that proved to be a vandal's pit. Many shallow little pits scattered over the rear of the cave--where no cultural material occurred--testified to repeated "tests" by relic hunters. The shallow pits in the back of the cave made it evident from the outset that the cave deposits were confined to the front third of the cavern, deep near the front and gradually feathering out toward the rear. As a result of two prior tests by the University of Utah, it was known that the site contained both paleontological and anthropological deposits of probable value to scholars, but the depth and extent of the deposits was not fully anticipated. Probably because it is remote, lying in difficult roadless terrain, the site proved to have suffered less from looters than had been feared. Neither had the cave walls been greatly marred by names and graffiti. The name of Lorin Wilson, 1893, was the only inscribed evidence of earlier visitors. Faint traces of red paint on the west wall may attest the onetime presence of painted pictographs. Due to the constant exfoliation of tiny spall and even individual sand grains of the wall, perhaps through thermal erosion, no pictograph outlines remained. What may be faint traces of incised rock art on the cliff near the portal were noted, but no zoomorphic forms or motifs could be de-

ciphered. [Note: It should be pointed out that a fictional compass was established in excavating the cave; the back of the cave was referred to as north, the portal as south, and the walls east and west, respectively. The mapping control system, then, though internally consistent, is in fact aligned some 45° west of magnetic north.]

STRATIGRAPHY

The stratigraphy of the site seemed bewilderingly complex when the first exploratory trench was sunk alongside the west wall at the front. After an initial period when the situation defied reasonable explanation, understanding finally dawned. What became very clear was that the cave's history divided itself into five episodes, called *units* in Jennings (1975), a usage continued here. Each unit consists of a sterile sand basalar component beneath a second component yielding prehistoric data of one kind or another. Unit I, as already mentioned, consisted of sand and herbivore dung. The base of the unit was clean pink sand and spall. Both sand and spall are presumed to have come from the walls and roof over the many millennia prior to the use of the cave as a resting place by megafauna. The dung of mammoth, bison, horse, camel, and sloth was recovered. The mammoth is further represented by the tips of two juvenile tusks, deemed to be juvenile because each lacks the chipping and blunting characteristic of the tusks of an adult elephant. One tusk tip lay *beneath* the dung *on* the underlying sterile sand at the plane of their contact.

From the dung itself, six scraps of bone were recovered. All that could be identified were bison--a mandible fragment, several teeth, a rib tip--except one proximal end of a tibia, which was probably elk. It was too small for bison, but well outside the usual range of deer. Although six sq. m. of the dung was removed with great care and subsequently screened, there is no ready explanation for the scarcity of bone or the

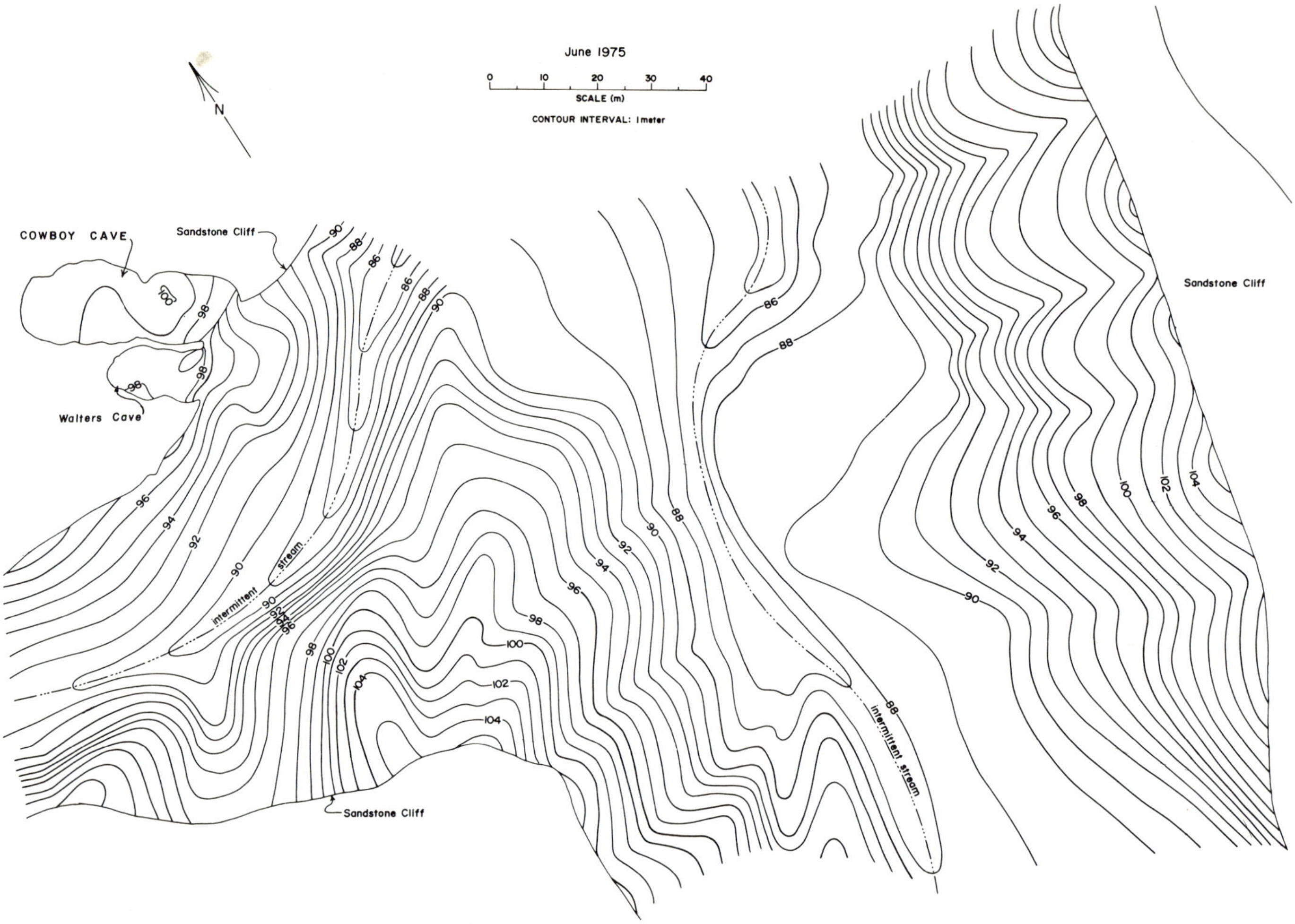

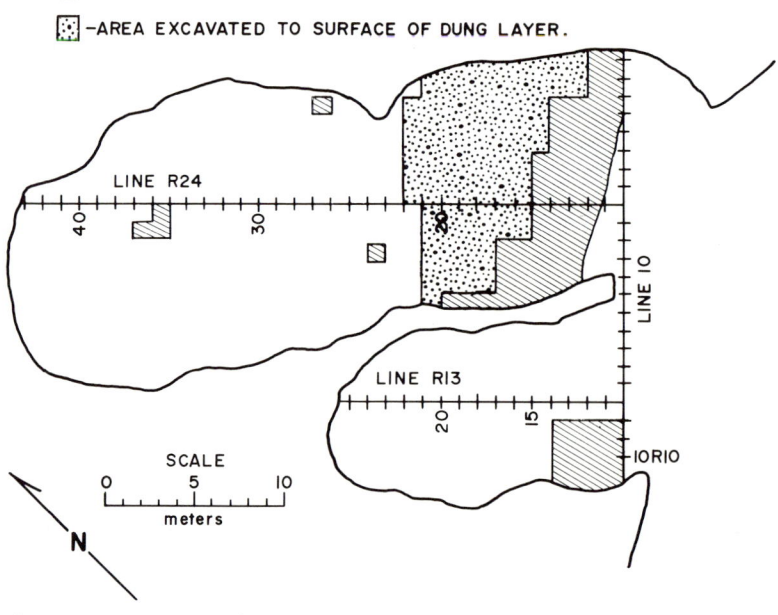

June 1975

SCALE (m)

CONTOUR INTERVAL: 1 meter

COWBOY CAVE

Sandstone Cliff

Walters Cave

Sandstone Cliff

Sandstone Cliff

intermittent stream

intermittent stream

FIG. 5.   Topographic map of canyon in front of Cowboy and Walters Caves.

⬜ —AREA EXCAVATED TO CAVE FLOOR.

⬜ —AREA EXCAVATED TO SURFACE OF DUNG LAYER.

LINE R24

40   30

LINE 10

LINE R13

20   15

10R10

SCALE

0   5   10

meters

N

FIG. 6.   Interior of two caves indicating area and controlled grid coordinates.

[10] AP 104/UU

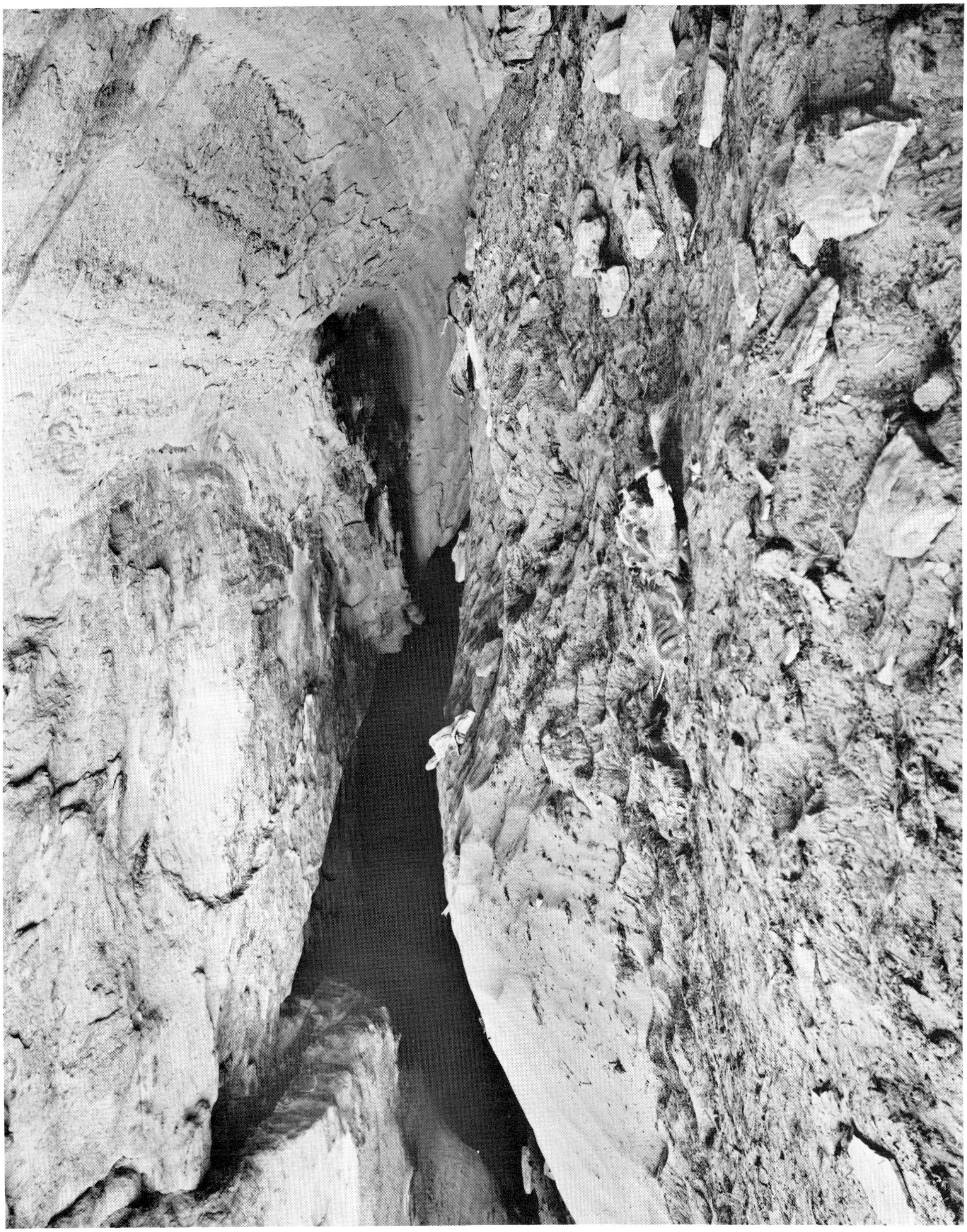

FIG. 7. East side of Cowboy Cave portal before excavation. Dessicated wild burro in foreground.

FIG. 8. View of interior mapping and initial exploratory trench.

FIG. 9. View toward protal from inside Cowboy Cave. The recent dune is prominently visible to the left of center.

FIG. 10. View toward portal from inside Walters Cave. Since occupation, large spalls occur more frequently than in Cowboy Cave. No significant dune formation had occurred here.

lack of whole bone. All pieces were broken and small; only the mandible shows a faint polish, as if its broken end had been used as a scraper or as a polishing device against some soft material. Hair picked from the dung deposits near the west wall was also identified as bison. Much of the dung was of finely ground grasses and was therefore ascribed to bison. It contained identifiable plant parts, as well as pollen. Grasses were dominant in two separate analyses, the species represented being *Sporobolus* (dropseed), *Oryzopsis* (Indian ricegrass), *Corispermum* (bugseed), and *Stipa* (needle and thread grass). (See *Appendices II* and *III*.)

The sterile sand beneath the dung, particularly near the walls, was lightly cemented with a carbonate. When cut and cleaned, it closely resembled caliche-- i.e., white streaks of concentrated mineral veining the clean sands. In view of our finding scattered gypsum crystals in the sand, it is presumed that the cementing occurred prior to 12,000 B.P. during a period wetter than now, when ground water containing gypsum salts moving in solution through the Navajo sandstone periodically moistened the basalar sands. Toward the central axis, the sands were more lightly cemented.

Above the dung, the deposits are cultural, all separated from each other by sand strata, as already mentioned. These are labeled Units II through V, V being the most recent (the strata within the units are designated Strata Va, IVc, etc.). The sand component of Unit II, as were all later sands, was extensive, but varied considerably in thickness over the site. On internal evidence, such as a few large spalls, some of Unit II sand is believed to have been deposited from the cavern ceiling, but some must have been wind deposited. It was not cemented to any perceptible degree; in fact, the field notes stress its lack of stability as compared to the sands of Unit I.

The sands of Unit II contained occasional flecks of charcoal and scat-

tered artifacts in the upper fraction. They are presumed to result from traffic during the first human use of the cavern directly on the surface of the sand. These sands are also notable for an extensive, included layer of leaves, sometimes as thick as 10 cm. Most were oak leaves from *Quercus gambeli*, a species still to be found near the cave. Twenty-three other species were represented among the other leaves. Both trees and shrubs identified are listed in *Appendices V* and *VI*. The extensive deposit of leaves, readily transported by wind, probably indicates that heavy thickets once stood directly in front of the cave. That assumption leads to the further supposition that there was more available moisture at that time (8900 B.P.) than now. The supposition is strengthened by the macrofossils and pollen in the dung at a much earlier date (see *Appendix II*). Similar strata of oak leaves are reported at other sites in the Canyonland province--e.g., Davis Kiva and Horse Shoe Alcove (Jennings 1966) as well as at DuPont Cavern (Nusbaum 1922). Interestingly, in all of the recorded cases the oak leaf deposits precede any evidence of human occupancy.

The cultural component of Unit II, while extending from wall to wall over most of the cave, was nowhere heavy. As Figure 11 shows, its greatest depth was never as much as 10 cm. It can best be described as a thin, ashy layer, with scattered lumps of charcoal and a few well-developed firezones or surface fireplaces, where ash and charcoal were more concentrated.

Upon Unit II a very thin but continuous layer of pink sand forms the *lower* component of the next unit, Unit III. The upper portion of Unit III is the most complicated and varied of all the cultural deposits. There were literally dozens of interleaving and overlapping short strata, local lensing, and truncated strata that resulted from frequent redeposition of the debris comprising the unit. Although the deposits at first bewildered the excavators, it was possible to distinguish the several substrata in the unit, and remove each as an entity

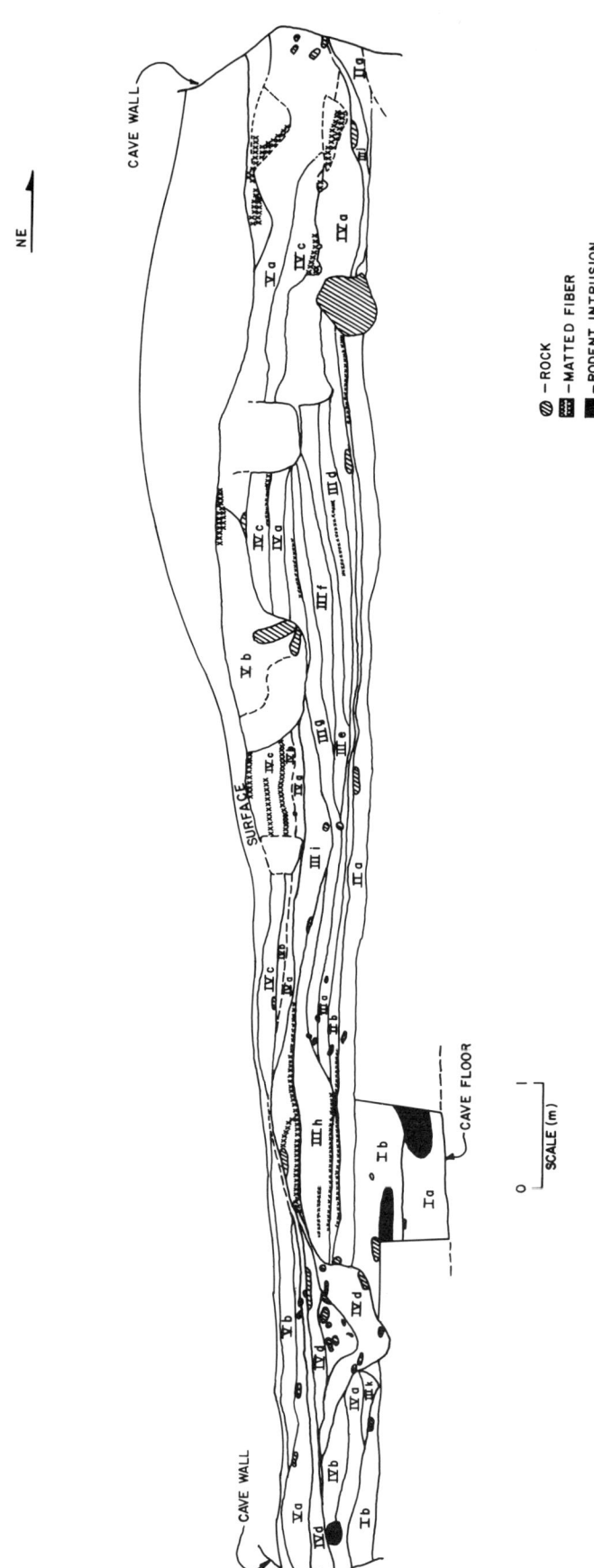

FIG. 11. East-west cross section (to scale) of the deposit. This is toward the rear of the cave about 1 m. beyond the point of greatest depth of cultural deposit.

that recorded an episode in the depositional history of the site. However,
the interpretation of the unit, once
arrived at, is simple and sensible.
Unit III consists of the debris generated by a long-term occupancy of the
cave; the preferred dwelling areas were
alongside the two walls. Large extensive hearth surfaces were distributed
between the ashy debris and the walls
for several meters. The fill is cred-
ited to frequent cleaning, by pushing or
scooping the ashes and other trash from
the cooking/living zone where the fires
were continuously used, to the center of
the cave. The fire areas had even been
dug or deepened through Unit II and into
the dung or upper component of Unit I.
The spoil from the scooping could actually be recognized in Unit III fill by
the presence of small lenses of redeposited dung. The fired areas along the
walls and into the dung are shown in Figure 11, an accurately scaled cross section of the deepest portion of the cave
fill. They are emphasized in the
schematic-interpretive cross section,
Figure 12. Therefore, when Unit III occupancy ended and human use of the cave
temporarily ceased, there were two long
ridges or "wind rows" of ashy midden
fill extending from the portal toward
the rear. The ridges paralleled the
walls; a valley or trough lay between
them. Between the ridges and the walls,
on both east and west sides, were the
hearth and habitation locations.

A description of the cultural fill
is difficult because the distinguishing
characteristics of any individual sub-
unit tended to rest on specific internal
attributes based on percentage of component material, compactness, or other
attributes instead of on marked differences in color, content, or other contrast with substrata above or below it.
The entire unit can therefore be described as numerous thin, dusty layers
of redeposited ash; with many lenses of
charcoal, also evidently redeposited
(i.e., with no basins, nor charred or
reddened zones in association); a high
but varying vegetal content; and scattered thin, local zones of mixed sand

and ashes. Often the gray ashy-charcoal
layers were separated from each other by
layers of almost pure vegetal chaff--
usually the straw and seed husks of
grasses. If living were concentrated
along the sides, then the ridges and the
trough between can be identified as work
areas or zones where grass seeds were
winnowed after harvesting. An occasional
small zone of shredded, matted cedar
bark, with no discernible function, was
also noted.

There is undoubtedly some earlier
material in the upper substrata of the
ridges of Unit III. The truncated strata
nearest the walls, as well as the internal evidence in the nature of the strata
themselves, assure this. The disturbance
should be minimal, however, if the concept of frequent cleaning of the house-
hold hearth areas is correct. Redeposition would be confined to the ridges
and would also affect only the later
strata. Those layers at the base of the
ridges and in the valley between show no
truncation. There was no way to determine exactly how extensive the reversal
of strata may have been. In any case,
the cultural importance of the reversal
is negligible because the Unit III occupancy lasted only about 800 to 900 years
(see Table 2). Only Unit III shows such
disturbance; the other cultural strata
in Units II, IV, and V are continuous,
showing no truncation or intentional
shifting. Contamination would be probable, therefore, only in Unit III. Many
local disturbances originating in Units
IV and V posed no problem of interpretation; they were pits, readily recognized, whose fill was segregated during
excavation, all care being exercised
during their cleaning to avoid contamination of depositional strata with fill
from intrusive pits from the later,
higher units.

Unit IV was separated from Unit III
by its unbroken lower sand element. This
was the thickest of all the sand layers,
and obscured any surface evidence of the
central valley or swale, being thickest
in the center of the cave. This unit,
like the other cultural layers above
Unit III, was continuous from wall to

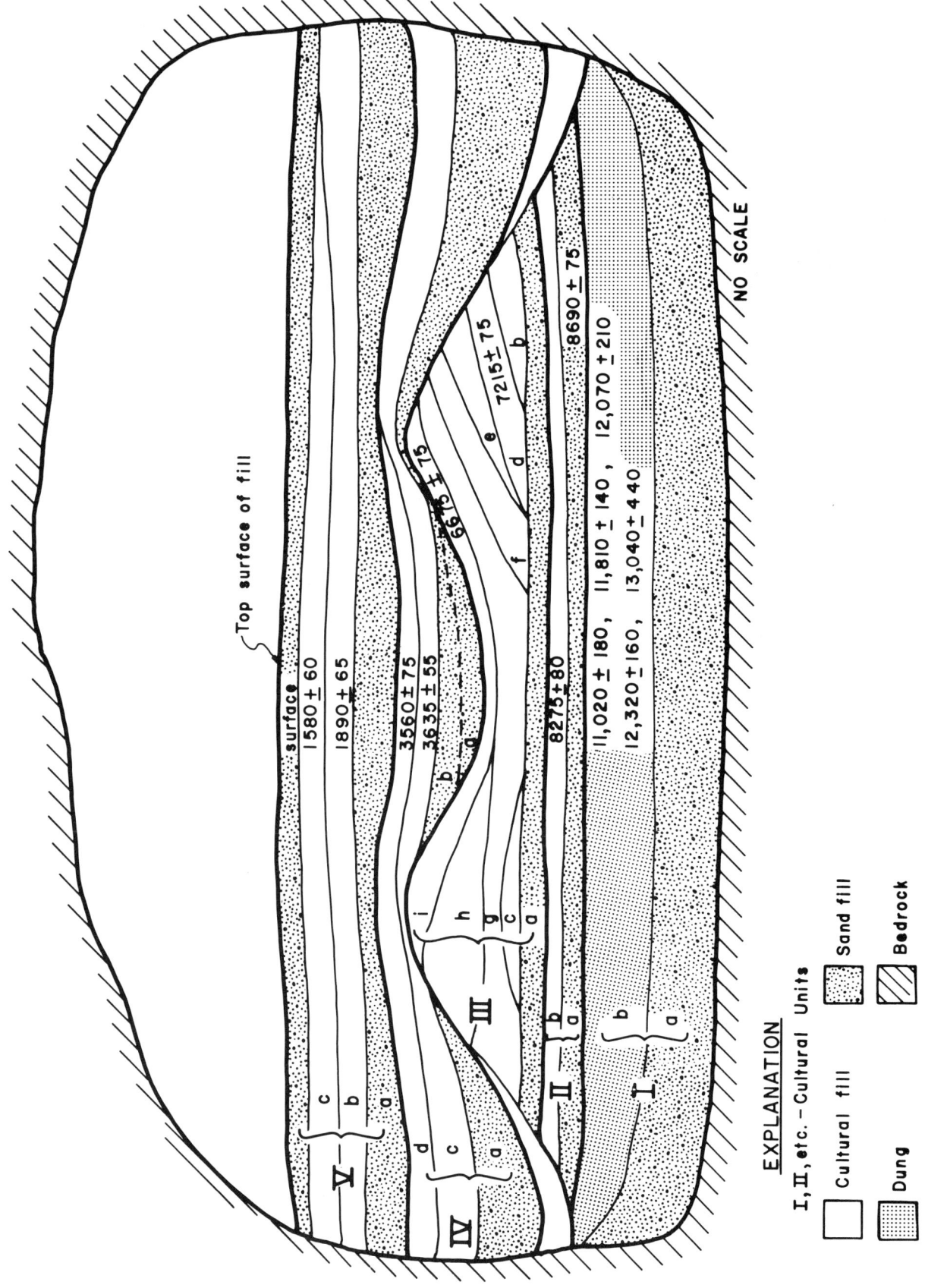

FIG. 12. Idealized cross section emphasizing substrata showing key radiocarbon ages. This figure should be studied in conjunction with Table 3.

TABLE 2

APPARENT CHRONOLOGY OF COWBOY CAVE MAJOR EPISODES

		I Dates as Given (Ignoring ± Range)			II Dates Using Extreme ± Ranges			
UNIT	Break (yrs)	B.P.	B.C./A.D.	Use Span (yrs)	B.P.	B.C./A.D.	Use Span (yrs)	Break (yrs)
V		1495±60 1890±65	A.D. 455 A.D. 60	395	1495−60=1435 1890+65=1955	A.D. 515 A.D. 5	510	
Sand	1440							1295
IV		3330±80 3635±55	1380 B.C. 1685 B.C.	305	3330−80=3250 3635+55=3690	1300 B.C. 1740 B.C.	440	
Sand	2755							2630
III		6390±70 7215±75	4440 B.C. 5265 B.C.	825	6390−70=6320 7215+75=7290	4370 B.C. 5340 B.C.	970	
Sand	1060							905
II		8275±80 8690±75 (−100)	6325 B.C. 6640 B.C.*	315	8275−80=8195 8690+75=8765 (−100)	6245 B.C. 6915 B.C.*	670	
Sand with leaf layer	2330							2070
I (Dung layer)		11020±180 13040±440	9070 B.C. 11090 B.C.		11020−180=10840 13040+440=13480	8890 B.C. 11530 B.C.		
Basal sand								

*Estimate using leaf layer date (IIa) minus 100 years.

AP 104/UU [19]

wall. No redeposition or shifting was detected. The description of the fill is about the same as for Unit III—varyingly gray, ashy, charcoal flecked, and somewhat sandy, with a heavy vegetal component indicating much winnowing activity. There were more numerous and larger areas of matted cedar bark. Again, no function could be ascribed to the cedar bark, although their large dimensions lead to the speculation that they were pallets or sleeping "nests." However, no basining or other shaping that one might expect in a deliberately fabricated nest or sleeping pad was noted. A few *in situ* surface fires were scattered through the layers. Charcoal was abundant.

It was evident that the space utilization practices were quite different from Unit III times. The household(s) were randomly dispersed over the entire western half of the cave, while dozens of (cache?) pits (usually lined with a thick cedar bark layer) and one or more metates—in the bottom or set on end leaning against the pit wall—pocked the eastern half of the cave. There, disturbances sank into the sand component of Unit IV. The pits often cut into earlier ones so that frequently a two- or three-pit sequence in one small area was observed. Figure 13 illustrates an example. Except for the bark and the metates, whole and fragmentary, the pits contained only ashy fill. More than 24 barklined pits originated in Unit IV fill.

The next to last episode in the occupancy of the cave is evidenced by Unit V. The cultural component is again sealed, separated from Unit IV by a continuous sand layer. Again, too, the cultural layer is continuous, being thickest to the east of the centerline of the cave.

Throughout the ashy-vegetal-sand fill are several surface fire areas, marked by ash and charcoal, with the chaff immediately beneath charred from the heat of the fire. In one or two instances there are large ashy areas, a meter or more in diameter, where the surface fire escaped and burned the underlying chaff for a depth of several centimeters.

The cedar-lined pits originating in Unit V were very numerous. Over 50 are recorded (Fig. 14). As a result certain areas are quite disturbed and we can only recognize that some of the fill in Unit V *must* be spoil dirt from the aboriginal digging. There would therefore be some contamination of Unit V from Unit IV. How much can only be speculated upon. We do know, however, that comparison of the artifacts recovered from undisturbed zones to the west of the center of the cave in Unit IV and in Unit V showed them to be quite similar despite a 1,000+ year break in occupation shown by radiocarbon dates. Thus we can conclude that the probable contamination did not seriously skew the artifact inventory or cultural interpretation of the two layers. The seeds and chaff associated with Unit V are chenopods instead of grass; this perhaps reflects cultural preference. More likely, however, it represents the cultivation of the amaranth.

Unrelated to, and postdating, Unit V was evidence of a brief reuse of the cave. The evidence consisted of a square, stone slab-lined box or cist surrounded by a floor or mat of grass some 3 m. in dia. The cist was chinked with "twists" or bundles of shredded cedar bark. Adobe plaster around the top of the slabs suggests that it had once had a stone cover sealed with adobe (see Fig. 15). The box was not large, being about 40 cm. on a side. No artifacts were found in the box or in the grass mat surrounding it. The box had evidently been opened by whoever dug the looters' pit or crater into the dune, as described elsewhere. The complex is interpreted as a storage cache, not involving actual residence in the cave. The makers could have been a transient party of Anasazi foragers or hunters. At least, the cist resembles those found in many caves and overhangs in southeastern Utah.

In summation, Cowboy Cave contained five major strata and one later, unnumbered phenomenon. The uppermost four—Units II through V—are cultural. All units include an extensive sterile zone

FIG. 13. Photograph of east-west cross section toward rear of cave. This picture is useful in several ways. In the foreground (in the dung layer) are two large spalls of which there were many in Unit I. Scattered through the cultural deposits smaller spalls can be observed. In Units IV and V, the spalls were rare and quite small. In the center, the east ridge of debris comprising Unit III is very clear. To the right is the deep, scooped-out living area, and, at the base of the section, much ash and heavy charred dung show plainly. Just left of center, originating in Unit V and intruding into Unit VI, is one of the many juniper bark-filled pits.

FIG. 14. View similar to Fig. 13 showing west ridge and edge of west scooped-out zone.

FIG. 15. Square, slab-lined cist a few centimeters beneath surface of dune. The cist was chinked with cedar bark bundles; a rather thick layer of loose grass extended out about 1.5 m. on all sides.

AP 104/UU [23]

TABLE 3

RADIOCARBON DATA

COWBOY CAVE

Lab No.	FS No.	Material Assayed	Radiocarbon Age			Unit or Stratum	Comment
			B.P.	B.C.	A.D.		
SI2425	1940	Charcoal	1495±60 B.P.		A.D. 455	Va	From pit (F183) in red wind-blown sand layer.
UGa1548	1517-1*	Corn	1555±70 B.P.		A.D. 395	Prob. V	Cached in skin bag.
SI2426	1683	Bark of *Juniperus* sp. and stalks of *Artemisia* cf. *dracunculus*	1580±60 B.P.		A.D. 370	Vc	Associated with semicircular arc of stones and small stone cist in stratum of Unit V, marking terminal occupation of the cave.
SI3012R**	1517*	Corn	1670±70 B.P.		A.D. 280	Prob. V	Cached in skin bag found in a shallow pit in ashy midden layer. dC13 estimated as -12.0%.
SI3172	1517-1*	Corn	1855±70 B.P.		A.D. 95		
SI2423**	1516	*Sporobolus* cf. *giganteus*	1840±65 B.P.		A.D. 110	Prob. V	From a fiber pad overlying the cache of shelled corn, FS1517. dC13 = -15.6%.
UGa1053		Charcoal	1890±65 B.P.		A.D. 69	Vb? (NP)	
SI2422	1517*	Corn	2075±70 B.P.	125 B.C.			Shelled corn cache. Same as UGa1548, SI3012R, and SI3172.
SI2495**		*Sporobolus cryptandrus*	3330±80 B.P.	1380 B.C.		NP	Grass from skin bag recovered in 1973 testing. dC13 = -15.6%. Probably from IVc.
SI2998	2293	Wood	3560±75 B.P.	1610 B.C.		IVd	
SI2715	1373	Charcoal	3635±55 B.P.	1685 B.C.		IVc	
SI2421	2158	Charcoal	6390±70 B.P.	4440 B.C.		IV? (or III)	Found in sterile red sand layer.

Lab No.	No.	Material	Age (B.P.)	Age (B.C.)	Unit	Comments
SI2420	485	Sandal fragment of *Yucca harrimaniae*	6675±75 B.P.	4725 B.C.	IIIi	From ashy midden fill. Expected to provide terminal date of occupation before deposition of overlying windblown sand of Unit IV. C13/C12 ratio of -24.3% suggests that this C.A.M. plant was harvested during cooler and moister season.
UGa637	13	Charred wood	6830±80 B.P.	4880 B.C.	III?	Recovered during preliminary testing.
SI2419	1154	Charcoal	7215±75 B.P.	5265 B.C.	IIId	From within midden fill of ash, charcoal, and matted fiber.
SI2418	508	Charcoal	8275±80 B.P.	6325 B.C.	IIb	In first thin cultural layer (F41) on surface of windblown sand of FS860 below.
SI2417	860	Leaves of *Quercus gambeli* and *Quercus* sp.	8690±75 B.P.	6740 B.C.	IIa	From sterile layer of pink windblown sand. Precultural.
A1660		Dung	11,020±180 B.P.	9070 B.C.	Ib	Collected by Univ. of Arizona associates.
UGa636	3	Dung	11,810±140 B.P.	9860 B.C.	Ib	Recovered during preliminary testing.
A1653		Dung	12,070±210 B.P.	10,120 B.C.	Ib	Collected by Univ. of Arizona associates.
A1800		Dung	12,320±160 B.P.	10,370 B.C.	Ib	Collected by Univ. of Arizona associates.
A1654		Dung	13,040±440 B.P.	11,090 B.C.	Ib	Collected by Univ. of Arizona associates.

JIM WALTERS CAVE

Lab No.	No.	Material	Age (B.P.)	Age (B.C.)	Unit	Comments
SI2416	370	Sandal	8875±125 B.P.	6925 B.C.	NP	

*Assays from same shelled corn cache.

**These ages are corrected for C13 fractionation.

of pink sand, sealing each cultural unit from those above and below it. The sand layers range in thickness from 2 to 3 cm. to over 50 cm. These represent interruptions in human use of the cave-- breaks also shown by the suite of radiocarbon dates (see Table 3). There is evidence of significant culture change between Unit III and Units IV and V. Thus, the typology of artifacts lends support to the chronology provided by radiocarbon analyses.

A word of caution is in order here. From the above paragraphs, it could be inferred that *no* artifacts occur in the sands that form stratum a and sometimes stratum b within each major unit, i.e., that the sands are sterile, lacking in artifacts. Such is not the case. Varyingly, artifacts were pushed down into the loose upper sand zones by foot traffic when occupancy of the cave was renewed.

EXCAVATION PROCEDURE

As mentioned, the cave fill was completely dry. No evidence was detected suggesting that any moisture had entered the cave after 12,000 B.P. Earlier, there had been the slight seepage recorded in the light cementation and the gypsum "caliche" of the sands on the cave floor.

Thus, given the ashy nature of the fill, it was obvious at the outset that excavation would be disagreeable work in that vision would be hindered and dust masks would be imperative. The reality exceeded expectation. Screening the fill at the lip of the cave produced a large volume of ashy dust, most of which seemed to come back into the cave as the strong southwesterly winds eddied and swirled in the cave mouth. However, the shaded cave interior remained quite cool--probably in the high 70's--so that high summer heats did not compound the discomfort the excavators experienced.

The procedures adopted for excavation were the more or less standard ones utilized with random, nonstructural deposits. After an arbitrary one-meter

grid was established over the site, each separable layer was removed as a unit over the one-meter square. This fill was screened, the lots of specimens were bagged separately, and each lot was given its permanent identifying number. This procedure was also followed with the dung. Sterile sand, however (especially Stratum Ia), was dumped over the side. Any fill which could not be quite firmly tied to a substratum or a square (as in the case of a sand slump, for example), or to a cultural phenomenon (such as a pit), was most often screened but its artifact yield was assigned to the "No Provenience" category. Sometimes, but rarely, uncontrolled fill went directly to the dump.

Digging and screening were done by two crews of four or five persons, directed by a sixth person charged with supervisory responsibility; this person was essentially a subforeman. The crews worked on opposite sides of the cave and their observations were made independently. Thus the numbers assigned to the strata were locally generated and differed between the "digs." As work proceeded it became possible to determine that all major units extended across the site, so that combining two or more numbered strata into one was possible. This was the responsibility of the director of the research, and went on continuously in the field, the combining being done on the basis of physical union of strata as the east and west trenches met along the median line.

Subsequent analysis of the data generated was done by the substrata that were recognized in the field as incremental units. The results were then combined into the more inclusive units previously described for the reporting that comprises this volume. Even though the results of the substratum studies provided tight, fine-grained controls, the detailed data are inflicted on the reader only when necessary to make one point or another. The records and tabulations documenting the detailed studies are preserved with the field records at the Utah Museum of Natural History. Table 4, provided here, correlates the major phenomena (features) with units and

TABLE 4

CORRELATION OF FIELD DESIGNATORS WITH
STRATA NUMBERS USED IN THIS REPORT

Stratum	Feature Numbers	Stratum	Feature Numbers
Sur	18, 21, 40, 61, 93, 95, 129, 134, 143, 148, 149, 152, 156	IIIk	29, 32, 67, 141, 151, 162
		IIIj	83, 89, 175, 178, 203
Vd	45, 49, 50, 52, 64, 73, 155	IIIi	77, 161
Vc	144, 145, 164, 185, 187, 191, 192, 198, 204	IIIh	36
		IIIg	78, 109, 182, 184
Vb	4, 33, 34, 94, 99, 116, 117, 118, 120, 153, 163, 180, 186, 190, 193, 194, 195, 197, 199, 200, 201, 206, 208, 209	IIIf	85, 98
		IIIe	103
		IIId	102
Va	7, 27, 100, 114, 119, 123, 124, 125, 126, 132, 133, 168, 172, 181, 183, 202, 210, 212	IIIc	165
		IIIb	171
IVd	28, 34, 112, 146, 147, 150, 154, 160	IIIa	86
		IIb	41, 140, 174
IVc	37, 68, 69, 106, 135, 136, 137, 157, 167, 188	IIa	14
IVb	31, 38, 142, 166, 173, 196, 205, 207, 211	Ib	15
		Ia	17
IVa	38, 80, 90, 97, 108, 110, 113, 115, 122, 127, 179, 189		

strata used in this report. This is provided to facilitate any future review or restudy of these data. Table 4 also shows the strata within the major units, as follows: Units I and II have only two subdivisions, a and b; in both cases, a represents the sand layer. Unit III strata run from a through k; a again is basal sand, with b through k being cultural. In Unit IV, a and b are sand; c and d are cultural. Unit V contains a (sand), and b, c, and d, which are cultural. Not shown are the cache pit and grass bed that evidence a brief post-occupancy reuse of a small area on the east side of the cave.

CHRONOLOGY

Although study of Cowboy Cave began with the knowledge that the cultural accumulation was bracketed by two radiocarbon year ascriptions (6830±80 B.P., UGa637 and 1890±65 B.P., UGa1053), and with one date for the dung deposit (11,810±140 B.P., UGa636), the controls on the samples were not fully integrated with the final controlling designations just described. The samples, collected at the time of test trenching, merely represented what appeared at that time to be the earliest and latest of the cultural layers, with the dung sample being a mix of samples from the lower and upper zones of that deposit.

After excavation, a suite of samples under good control was submitted to the Smithsonian Institution Radiocarbon Laboratory. The earlier dates, UGa636 and UGa637, are congruent and are integrated into Table 3 in their proper chronological position with respect to the Smithsonian dates and other later dates from the Geochronology Laboratory at the University of Georgia. The University of Arizona laboratory also dated some material from Cowboy Cave; these dates, too, are accordant with the Georgia and Smithsonian results. All radiocarbon year assays appear in Table 3.

Although both Tables 2 and 3 appear at first glance to be straightforward, some explanation of Table 3, which summarizes the radiocarbon data, is required. Here, all the assays are included even though, in one or two cases, the provenience of the sample is uncertain. One such date is SI2495, derived from grass found in a fawn's head skin bag. On the basis of our observations during excavation, we concluded that the fossil packrat nest where the bag was found was to be correlated with Unit IV, a correlation made two years after recovery of the specimen. The date of 3300 B.P. is entirely compatible with the age of other controlled samples from Unit IV; accordingly, with the radiocarbon dates supporting the field conclusion, the object is assigned to Unit IV.

Similarly, the dates SI2422, SI2423, SI3012, SI3012R, SI3172, and UGa1548 all refer to a cache of shelled corn and an associated mat of grass which our field notes assign to Unit IV (Fig. 16). The corn dates all read *less* than 2,000 years, well within the range for other samples from Unit V, and more than 1,300 years later than any date in Unit IV. Therefore, the corn is ascribed to the later Unit V. In doing this we were led to the unpalatable conclusion that the corn cache was in an unobserved intrusive pit from Unit V *into* Unit IV.

Another possible age anomaly is the apparent association of the large clay figurine from Walters Cave with a sandal dated 8875 B.P. (SI2416). The sandal age ascription is probably correct, but the association with the clay figurine *may* be spurious. Although a few figurines, a loaf-shaped object, a cone, and several clay fragments came from Units II and III, most of the unfired clay objects are restricted to Units IV and V. Even with the figurine provenience clouded, but possibly correct, the sandal age is compatible with the archeological evidence from Units II and III. That the Walters Cave sandal yields the greatest radiocarbon age of any sample suggests that the occupancy of Walters Cave preceded the use of Cowboy by a few hundred years. As indicated above, Walters Cave was thoroughly vandalized, leaving now no way to verify the possibility of its earlier use.

After the construction of Table 3, Table 2 was developed. It serves to indicate the duration of the cultural occupation of the four units, and the length of the breaks between occupations. Using radiocarbon figures without the ± variations, Units II, IV, and V appear to be very short lived--from 300 to 400 years in length. Unit III, the deepest cultural layer, lasted some 800 years. The breaks between occupancies all exceed 1,000 years. Using the extreme ± figures, the occupancy spans are in some degree lengthened with accompanying slight reduction in the duration of the breaks. Throughout the text only the read-out figures are cited.

FIG. 16. Two skin bags containing shelled corn. The substrata of Unit IV run undisturbed across the few centimeters of the bags that are still imbedded in the working face. The bags themselves were in a shallow pit lined, as can be seen, with cedar bark and grass. It was covered with a mat of *Sporobolis* grass. The text discusses the contradictory evidence of the excavation notes and this photograph and the much later radiocarbon date ascribed to the corn.

CHIPPED STONE
PROJECTILE POINTS

Richard N. Holmer

INTRODUCTION

The projectile point assemblage recovered from Cowboy and Walters Caves is typologically consistent with other sites of similar age in the eastern Great Basin and the Colorado Plateau. The tight stratigraphic control maintained during excavation yielded a clearly defined progression of several widely known point types, and the subsequent radiocarbon dates confirmed and refined their temporal distribution. In addition, the data from the two caves, in conjunction with similar data from Sudden Shelter (Jennings et al. In press), provided a valuable resource for auxiliary studies such as tool use [Holmer 1976b] and the cross-dating of other sites [Schroedl 1976].

The method used in the typing of the projectile points follows the one developed for use with the Sudden Shelter points (Jennings et al. In press). The method is entirely objective, based on a statistical analysis of morphology. A discussion of the techniques can be found in the Sudden Shelter report, and a more complete description appears in Holmer [1978]. A summary is provided here.

A discriminant analysis computer program (Nie et al. 1975) was used to analyze the shapes of projectile points from measurements taken from photo-graphs and drawings published in original site reports. The same measurements were taken for the Cowboy Cave points and introduced into the program as unknown types. The program provided statistics indicating which known type the point most closely resembled, and the strength of that resemblance. New types are derived from identifying clusters of points that do not fall within the range of any of the known types. In the case of Cowboy and Walters Caves, 95 percent of the collection falls within the ranges of four widely recognized projectile point types or groups: the Elko series, Northern Side-notched, Gypsum, and Rose Springs. Because the numerous measurements taken require detailed explanation, they are not included in this report, but can be consulted in Holmer [1978].

PROJECTILE POINTS

There were 138 whole and fragmentary projectile points recovered from Cowboy and Walters Caves. Of them, 121 are complete enough for identification as to type, although only 93 (77 percent of the total identifiable collection) are positively provenienced. Unless otherwise specified, this discussion applies to points from both caves. The following type descriptions are traditional in scope; the objective (metric) data are included in Table 5 and the provenience

TABLE 5

PROJECTILE POINT DIMENSIONS

Projectile Point Type	Length mm.	(N)	Width mm.	(N)
Elko Corner-notched	41.25± 7.78	(22)	22.26±4.21	(35)
Elko Side-notched	45.88± 7.81	(5)	22.00±3.56	(8)
Elko Eared	48.00±18.38	(2)	19.33±2.31	(3)
Northern Side-notched	49.38±11.46	(8)	24.4 ±3.37	(9)
Gypsum	49.41± 8.77	(22)	20.65±3.37	(36)
Rose Springs	25.33± 4.14	(12)	11.25±0.75	(18)

information in Table 6.

Elko Series

The Elko series has traditionally been divided into three subtypes: Corner-notched, Side-notched, and Eared. It has previously been noted [Holmer 1976a] that the Corner-notched and Side-notched are not two distinct groups, but are a continuum from one extreme to the other. A cut-off point, therefore, is completely arbitrary and cannot be defended statistically, but may be of some use for comparative purposes. The cut-off point established for the Sudden Shelter projectile points is used here.

Elko Corner-notched: These triangular blades have straight to slightly convex edges (Fig. 17t, v, w). Corner notches form tangs and an expanding stem that is narrower at its base than the maximum blade width. The base ranges from slightly concave to slightly convex.

Elko Side-notched: These blades are similar in form to the Elko Corner-notched, except that the maximum stem width is approximately equal to the maximum blade width (Fig. 17x, y, z, aa). Tangs are rarely present; the distal notch angle often approaches horizontal, giving a shouldered appearance.

Elko Eared: These blades are simi-lar in form to the Elko Corner-notched except that the base is markedly concave, forming a notch that results in an eared or bilobed stem (Fig. 17u).

Northern Side-notched

These triangular blade forms (Fig. 17o-s) have slightly convex edges. Horizontal notches are moderately high on the sides, forming a slightly contracting stem that is approximately the same width as the blade. The base is consistently concave, and the edges of many of the blades (seven out of nine) are markedly serrated.

Gypsum

These triangular blade forms (Fig. 17g-n) have convex edges. Wide corner notches form roughly square shoulders and a contracting, convex edged stem. A few specimens (6 out of a total of 37) have serrated edges; many points have remnants of pitch coating the stem and portions of the lower blade.

Rose Springs

These triangular blade forms (Fig. 17a-f) have straight to slightly convex edges. Corner notches form shoulders from square to slightly tanged, and stems from parallel-sided to slightly expanding. The Rose Spring points are narrow and

FIG. 17. Projectile points: a-f, Rose Springs; g-n, Gypsum; o-s, Northern Side-notched; t-aa, Elko.

TABLE 6

DISTRIBUTION OF PROJECTILE POINTS

	Elko Corner-notched	Elko Side-notched	Elko Eared	Northern Side-notched	Gypsum	Sudden (?)	Rocker Side-notched	Rose Springs	Desert Side-notched	Cottonwood Triangular	Fragments	TOTAL	Approximate Dates B.P.
Cowboy Cave													
NP	5		1	2	8	1	2	2		1	2	24	
TT2					1			1	1			3	
Unit I													
a													
b													
Unit II													
a													8700
b											1	1	8300
Unit III													
a	1											1	
b													
c													
d				1								1	7200
e													
f	1	1										2	
g											1	1	
h				1							1	2	
i				1								1	6700
j	1			3								4	
k				1								1	
Unit IV													
a	2				4						2	8	6400
b	3				1							4	
c	6	1			7						4	18	3600
d	2		2		3						1	8	
Unit V													
a	5				6				1			12	1500
b	4	2			5				4		4	19	

TABLE 6 (continued)

DISTRIBUTION OF PROJECTILE POINTS

	Elko Corner-notched	Elko Side-notched	Elko Eared	Northern Side-notched	Gypsum	Sudden (?)	Rocker Side-notched	Rose Springs	Desert Side-notched	Cottonwood Triangular	Fragments	TOTAL	Approximate Dates B.P.
c	1	1						3		1		6	1600
d													
Sur	4	2						4				10	
TOTAL	35	7	3	9	35	1	2	15	1	2	16	126	
Walters Cave													
NP	2				1							3	
TT								1				1	
Unit I													
A													
B													
Unit II													
A													
B													
Unit III													
A													
B													
C													
D		1										1	8900
Unit IV													
A													
B													
Unit V													
A	1							1			1	3	
B	1							3				4	
Sur													
TOTAL	4	1			1			5			1	12	

Total points--138 (100%). Total identified points--121 (88%). Total points in provenience--107 (78%). Total identified points in provenience--93 (67%).

thin compared to their length, and are considerably smaller than the types previously described.

Miscellaneous Projectile Points

Several projectile points of types not mentioned above were recovered. They are represented by only one or two examples, or their provenience is unknown. These are the Desert Sidenotched, Cottonwood Triangular, and three larger points that have their closest affinities to the Sudden Sidenotched and Rocker Side-notched reported for Sudden Shelter (Jennings et al. In press).

DISCUSSION

With a few minor exceptions the temporal distribution of projectile points at Cowboy Cave is congruent with the distribution of the same types at other sites in the eastern Great Basin and the western Colorado Plateau (Fig. 18). The earliest and most abundant point type found at Cowboy Cave is the Elko series, representing over 40 percent of the total identifiable collection, and occurring from approximately 8100 to 1400 B.P. The earliest occurrence of the Elko series at Sudden Shelter is similar to that at Cowboy Cave, although their termination at Sudden Shelter is considerably earlier (4700 B.P.) (Jennings et al. In press). The date range at O'Malley Shelter (Fowler et al. 1973) is from over 7000 B.P. until the Fremont occupation after A.D. 1. The Hogup and Danger sequences (Aikens 1970) are of a similar time range as O'Malley Shelter, suggesting use from over 7000 B.P. to relatively recently. Other eastern Great Basin sites also support an extended period of use for the Elko series points, suggesting they span the Archaic and Fremont occupations of the area.

The Northern Side-notched projectile points comprise approximately 7 percent of the total identifiable collection. They range in time from approximately 7200 to 6400 B.P. Both Danger and Hogup

Caves (Aikens 1970) demonstrate longer sequences, spanning a time period from 8000 to 1000 B.P. As has been previously mentioned (Jennings et al. In press), the sequences for both Danger and Hogup Caves are obscured by Aiken's (1970) misclassification at those sites of several point types, including the Northern Sidenotched. Sites in the northern Great Basin (Layton 1972, Swanson et al. 1964, Gruhn 1961) show that the period from about 7000 to 6500 B.P. marks the greatest use of the Northern Side-notched points. These data, in conjunction with the Sudden Shelter and Cowboy Cave data, suggest that the Northern Side-notched point type may be of some use as a time marker when found in significant numbers.

The Gypsum point comprises approximately 30 percent of the total identifiable collection from Cowboy and Walters Caves. Their occurrence begins abruptly at the bottom of Unit IV, with the concurrent disappearance of the Northern Side-notched points at the top of Unit III. A 2,830 year hiatus in cave occupation between Units III and IV was postulated earlier in this volume, based on stratigraphic and radiocarbon date interpretation. The existence of the hiatus is strongly supported by the projectile point sequence as dated at other sites in the region (Fig. 19). At Sudden Shelter (Jennings et al. In press), Gypsum points are introduced at approximately 4600 B.P. and continue until the strata terminate at approximately 3300 B.P. The dates reported for Gypsum Cave (Heizer and Berger 1970:17; Shutler 1967: 306) indicate a time range for the Gypsum point encompassing a period from 3000 to 2000 B.P. O'Malley Shelter suggests a considerably longer time span, although most of the points occur from 4000 to 2000 B.P. All the evidence suggests that the earliest possible date for the introduction of Gypsum points at Cowboy Cave would be about 4600 B.P., a date that falls within the break between Units III and IV. Additional support for a widespread hiatus comes from Fowler's (et al. 1973:71-72) conclusion that O'Malley Shelter was abandoned between 6500 and 4600 B.P., coinciding with the Cowboy

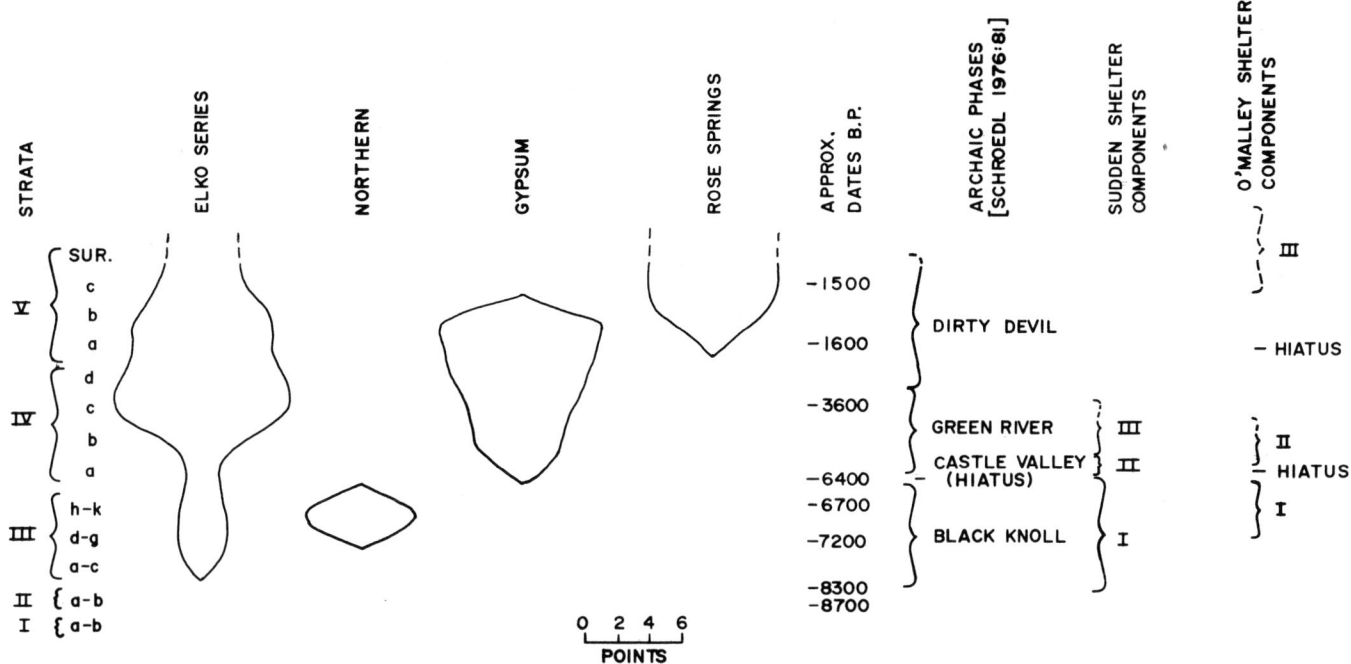

FIG. 18. Stratigraphic distribution of projectile points from the eastern Great Basin and western Colorado Plateau (two unit moving average), raw frequencies.

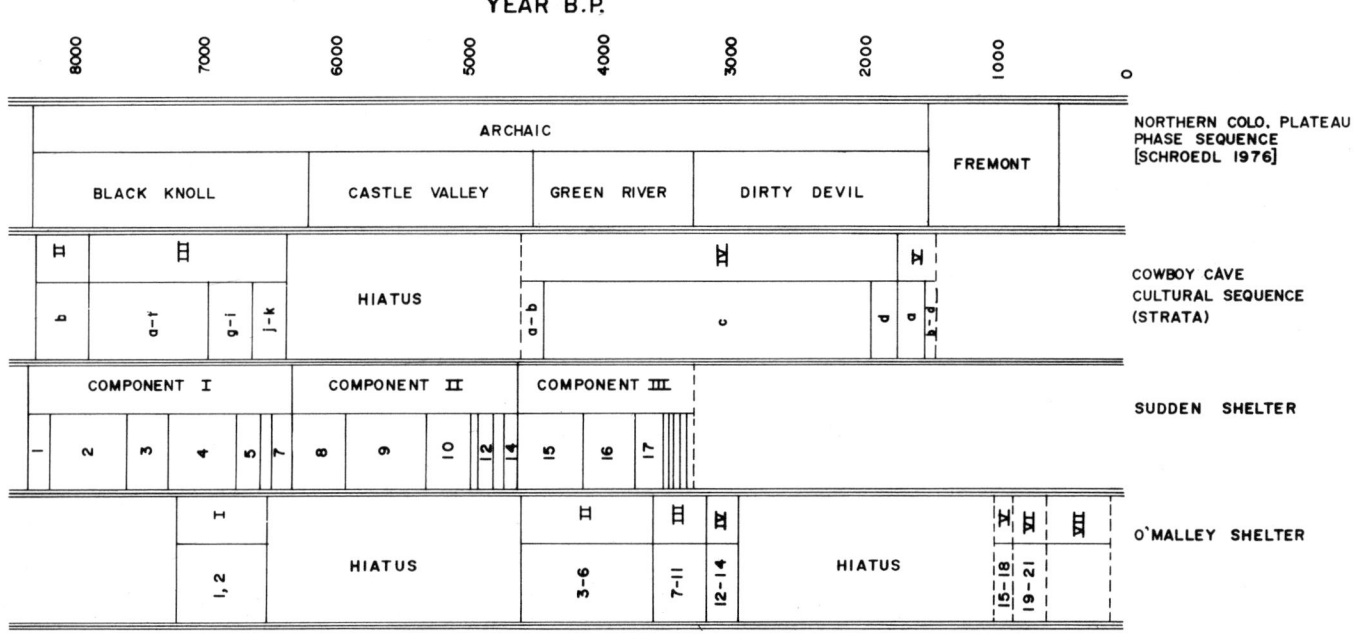

FIG. 19. Diagrammatic comparison of occupation spans, Northern Colorado Plateau, Cowboy Cave, Sudden Shelter, and O'Malley Shelter.

Cave break, and Jennings' et al. (In press) conclusion that Sudden Shelter was used significantly less during those years than in earlier or later periods.

Gypsum points may also be of interest in understanding the technological change that lead to the adoption of notchless points. Pitch is not evident on any points that occur earlier than the Gypsum points, indicating either that no pitch was used in the securing of the point to the shaft, or that pitch was not preserved from the earlier strata. Since basketry from the earlier strata did survive (see *Fiber Artifacts*), it seems likely that pitch would; the absence of pitch therefore suggests that pitch was probably not used to bind points to shafts earlier than approximately 3600 B.P.

With the discovery that pitch could be used as a binding agent, the need for deep side- and corner-notches was probably reduced for dart points. Deep notching probably remained a necessity for knife blades because of the lateral direction of force (as opposed to the longitudinal force at the tip of a dart point). The advantage of using pitch as a binding agent might be the increased ease of changing points. When a point is broken on a useable shaft, the pitch need only be melted and a new point inserted. With a sinew binding, the notched point would have to be cut free or unbound and the new point bound in place.

The Rose Springs projectile points are considerably smaller than all the previously discussed points and appear to mark a technological change at Cowboy Cave from the use of atlatl and dart to bow and arrow. Approximately 16 percent of the total identifiable collection are Rose Springs points. Their occurrence is limited to Unit V, beginning about 1600 B.P. and continuing to an unknown date. Hester and Heizer (1973:7-8) report the usual occurrences of Rose Springs points between 1400 and 800 B.P., even though other earlier examples have been noted (for example, Aikens 1970 reports the points at ca. 2500 B.P.). Hester and Heizer conclude that

the bow and arrow was probably introduced into the Great Basin at approximately 1450 B.P. or shortly thereafter. The Cowboy Cave information might make a date closer to 1650 B.P. more likely, or it may indicate that the bow and arrow was in use in the Colorado Plateau shortly before its introduction into the Great Basin.

The fact that no Pinto points were recovered from Cowboy and Walters Caves may be of some significance. A factor analysis of the remains of Sudden Shelter suggested a strong association between the Pinto points and the presence of deer bone [Holmer 1976b] that may indicate Pinto points were used for the hunting or butchering of deer. This hypothesis is not necessarily new as Aikens (1970:189) suggests the same relationship at Hogup Cave. The Cowboy Cave evidence is negative in that no Pinto points occur in the 8000 to 6400 B.P. date range where they might be expected; deer bone does not occur there either (see *Bone and Shell Material*). This implies that the site was probably not used as a hunting camp and the peoples who visited the site may not have been equipped for deer hunting and butchering. Also, the points that do occur during that time period (the Elko series and the Northern Side-notched) may have functions not restricted to hunting and butchering of large game. They may have been general purpose knives used, among other things, in the manufacture of other tools.

Another correlation observed in the analysis of Sudden Shelter remains [Holmer 1976b] that seems also to hold true for Cowboy Cave is between Gypsum points and bighorn sheep bones. At Cowboy Cave, neither occurs earlier than Stratum IVa. This may indicate that after the 6400 to 3600 B.P. hiatus people at Cowboy Cave hunted some large game. After approximately 1600 B.P., when the bow and arrow became prevalent, large game hunting continued using the Rose Springs points.

LITHIC ARTIFACTS AND DEBITAGE

Dennis G. Weder

INTRODUCTION

Several categories of lithic artifacts were recovered from Cowboy Cave and Walters Cave (see Tables 7 and 8). Since Walters Cave was only partially excavated, the major emphasis of this report will be on the artifacts recovered from Cowboy Cave.

Most of the lithic artifacts (69 percent from both caves) were functionally classified as cutting tools and subdivided into bifaces and utilized flakes. Microscopic examination of the cutting tools revealed that the bifacially flaked tools, characterized by edge-wear polish, were used almost exclusively for cutting soft materials. The utilized flakes, on the other hand, were characterized by use retouch, and thus were probably used primarily for cutting hard materials.

Microscopic edge-wear examinations were also performed on all of the projectile points. Wear polish indistinguishable from that on the bifaces was also found to be characteristic of 40 to 60 percent of the larger projectile points. With one exception, the smaller projectile points, classified here as arrowpoints, did not exhibit any wear polish.

Evidence of the aboriginal use of heat treatment of lithic tools was found in nearly all strata of Cowboy Cave.

Biface Manufacturing Sequence

The bifacially flaked specimens from Cowboy Cave were classified into three categories on the basis of the type of knapping flake scars they exhibited. Bifaces exhibiting predominantly hard-hammer knapping were classified as blanks. Preforms were those specimens that exhibited mostly soft-hammer knapping, and the knives were characterized by pressure retouch. The use of knapping techniques to classify bifaces follows Muto's (1971) procedure.

The use of the blank, preform, and knife terminology implies that a manufacturing sequence is being described. Microscopic studies of all bifaces revealed that wear polish was more prevalent on bifaces which were further along the manufacturing sequence. Of all the blanks, 18 percent exhibited wear polish, among the preforms, 38 percent had wear polish, and 69 percent of the knives exhibited wear polish. The large jump in wear polish prevalence from preforms to knives indicates that the wear polish on blanks and preforms was probably the result of opportunistic use.

The subtractive nature of knapping also supplied evidence of the blank-preform-knife manufacturing sequence of these artifacts. The dimensions and the minimum and maximum edge angles of all bifaces were measured and the averages for each category are shown in Table 9.

TABLE 7

DISTRIBUTION OF LITHIC TOOLS, COWBOY CAVE

	Test Trenches					Unit: I		II		III											IV				V				Sur	TOTAL
Stratum:	NP	1	2	3	4	a	b	a	b	a	b	c	d	e	f	g	h	i	j	k	a	b	c	d	a	b	c	d	Sur	TOTAL
Core	5								2			1				1		1			3	1	8		3	7	1		2	35
Blank	18	1							19						2	1	1	1			5	2	17	9	8	8			3	95
Preform	10	3							2			1	1		2	2	2	3		2	2	2	13	7	19	11	5	2	7	96
Knife	10														2		1	1	1		3	1	5			6	5	1	8	44
Utilized Flake	26	6						1	1				3		5			3	6	2	2		18	3	10	16	8	3	11	124
Scraper	4	1							2												2	3	3	2	2	3	1		5	28
Drill	1															1	1										1		1	5
Hammerstone	11							1					1			2	3	3		4		6			1	10	3		11	56
Projectile Point	3												1		2	1					1			1						9
Burin/graver/awl																		1	1				2						2	6
Chopper	1																						1		1	3			2	6
Grinder	2																									1			1	4
Crenated Cobbles		2											1										2	3		5	1		3	16
TOTAL	91	13						2	26	5	1	4	2	13	7	5	15	10		21	6		75	29	42	70	23	10	54	524
Debitage (No. of Flakes)	233	1				5	6	135	783	108	2	9	49	22	80	70	63	90	55	7	245	132	714	327	919	602	127	100	331	5,215

TABLE 8

DISTRIBUTION OF LITHIC TOOLS, WALTERS CAVE

			Unit I		Unit II		Unit III					Unit IV		Unit V			
Stratum:	NP	TT	A	B	A	B	A	B	C	D	K	A	B	A	B	Sur	TOTAL
Core														1			1
Blank	1												2		1		4
Preform		2									1	1	2	6	2	1	15
Knife													1	1		1	3
Utilized Flake	1	3				1					2	2		2	1		12
Scraper															1		1
Drill													1				1
Hammerstone													1				1
Projectile Point					1						1					1	3
Chopper		1								1		1					3
Crenated Cobbles	1											1		1			3
TOTAL	3	6			1	1				1	4	5	7	11	5	3	47
Debitage (No. of Flakes)	100		6	3	12					104	7	100	212	359	124	19	1,046

AP 104/UU [41]

TABLE 9

COMPARISON OF BLANK, PREFORM, KNIFE, AND DART-POINT DIMENSIONS AND EDGE ANGLES (ALL STRATA)

Thickness (cm.)

	Blank	Preform	Knife	Elko	Gypsum	Northern
Mean	0.96	0.66	0.55	0.47	0.52	0.52
Standard Dev.	0.35	0.23	0.13	0.06	0.07	0.11
No. of Specimens	90	80	70	49	38	9

Width (cm.)

	Blank	Preform	Knife	Elko	Gypsum	Northern
Mean	3.8	3.0	2.3	2.24	2.11	2.64
Standard Dev.	0.71	0.91	0.67	0.43	0.31	0.29
No. of Specimens	45	46	50	33	33	8

Length (cm.)

	Blank	Preform	Knife	Elko	Gypsum	Northern
Mean	5.85	6.4	4.8	4.21	4.82	4.60
Standard Dev.	0.86	1.9	1.3	0.98	0.88	0.85
No. of Specimens	23	6	22	16	20	8

Edge Angle (Degrees)

	Blank Min-Max	Preform Min-Max	Knife Min-Max	Elko Min-Max	Gypsum Min-Max	Northern Min-Max
Mean	45-70	40-52	40-44	40-44	37-43	41-44
Standard Dev.	8.9-11.5	8.2-11.4	9.3-9.7	6.8-10.0	6.8-8.1	9.2-9.2
No. of Specimens	94	85	70	38	36	9
Edge Angle Range	25	12	4	4	6	3

All three dimensions decrease from blank to preform to knife as would be expected. The edge-angle range also decreases from blank to preform to knife, indicating a refinement of the intended cutting edge.

The shapes of the bifaces also can be related to the manufacturing sequence. Oval bifaces were more prevalent in the earlier stages of manufacture. Of the blanks, 51 percent were oval while 29 percent of the preforms were oval, and only 20 percent of the knives. More specialized shapes, such as triangular and lanceolate forms, were present in low frequencies among the blanks (2 percent and 4 percent), rising to 12 percent and 13 percent among the preforms, and 24 percent among the knives.

The manufacturing techniques used to make bifaces did not exhibit any changes over the various periods of occupancy of Cowboy Cave. In addition, the size and morphological characteristics, such as shape and edge angle, showed no temporal trends.

The few biface specimens from Walters Cave conform in all characteristics to those recovered from Cowboy Cave.

Since such a high percentage (75 percent) of the bifaces recovered from Cowboy Cave were broken, the broken specimens were examined to determine whether the breakage was a circumstance related to manufacturing mistakes or utilization accidents. The examination revealed that both alternatives were present; some of the broken bifaces had wear polish and some did not.

Projectile Points

The projectile points from both Cowboy and Walters Caves were examined for edge-wear polish and the results are shown in Table 10. It is immediately obvious that the projectile points can be classified into two groups on the basis of size and the presence of wear polish. The larger point types, i.e., Elko, Gypsum, Northern, Sudden Side-notched and Rockerbase, exhibit wear polish in about 40 percent to 60 percent of the specimens. The smaller point types, Rose Springs, Cottonwood Triangular and Desert Side-notched, with one exception (shown in Table 8) typically lack any evidence of wear polish. In the subsequent discussion, then, a *dart point* refers to a larger point that may exhibit wear polish; an *arrowpoint* to a smaller point that probably lacks wear polish.

Edge angle and thickness measurements also verified the dart-point and arrow-point categories. The data resulting from these measurements are shown in Table 11. The edge-angle data show that the arrowpoints had smaller edge angles and smaller edge-angle ranges than the dart points, although the standard deviations reveal considerable overlap between the two categories. The thickness data, however, show a distinct difference between dart points and arrowpoints. The thickness populations do not overlap within one standard deviation of their respective means.

Knapping characteristics also can be used to distinguish between dart points and arrowpoints. Out of 24 arrowpoints recovered, in 18 examples the bifacial pressure retouch did not extend completely across the face of the arrowpoint. The incomplete bifacial retouch indicates that the arrowpoints were manufactured directly from small flakes.

The dart points, on the other hand, were all pressure retouched on both sides. In addition, the dimensions and edge angles of the dart points are very similar to the corresponding knife characteristics (see Table 7). In all cases, except for the mean width of Northern dart points, the thicknesses, lengths, and widths of the dart points are less than the corresponding knife dimensions. Also, the standard deviations of the dart-point dimensions are smaller than the standard deviations for the blanks, preforms, or knives. It seems probable that the dart points are at the most refined end of the subtractive biface-manufacturing sequence. The much smaller range of dart-point dimensional variation implies that the dart points were also more standardized tools than the knives. The degree of tool refinement is also shown by the edge-angle ranges, which, for the dart points, are less than or equal to that for the knives.

TABLE 10

WEAR POLISH ON PROJECTILE POINT LATERAL EDGES
COWBOY CAVE AND WALTERS CAVE

Projectile Point Type	Total No. of Specimens	No. of Specimens With Greater Than Very Slight Wear Polish	Percent of Specimens With Greater Than Very Slight Wear Polish
Dart Points			
Elko	39	23	59%
Gypsum	36	19	53
Northern	9	4	44
Sudden Side-notched	1	1	100
Rockerbase Points	2	1	50
Arrowpoints			
Rose Springs	19	1	5
Cottonwood Triangular	3	0	0
Desert Side-notched	1	0	0

It is possible that either a preform or a knife might have been pressure retouched appropriately to make the dart point.

The implications of the preceding discussion are several. First, the dart points, regardless of style, were apparently used opportunistically but regularly as cutting tools. The degree of wear polish on many of the dart points was greater than that found on any of the knives, and yet there was no apparent difference between the characteristics of the wear on the dart points compared to the knives. If the dart points had been mounted on an atlatl foreshaft, they could have conveniently served a double purpose as projectile points and hafted knives, thereby re-ducing the tool kit necessary for hunting trips.

The arrowpoints, on the other hand, were considerably thinner and smaller overall than the dart points and would therefore have been more fragile and less serviceable as cutting tools. It should not be ruled out that the arrowpoints were possibly used occasionally as cutting tools. One arrowpoint did exhibit definite wear polish. Due to the thinness of an arrowpoint, however, its lifespan as a cutting tool would probably have been rather short, thereby reducing the likelihood of its developing noticeable wear polish. It should be noted that the dart points, with and without wear polish, continued to be used even after the arrowpoints appeared in the archeological

TABLE 11

EDGE ANGLE AND THICKNESS DATA FOR DART POINTS AND ARROWPOINTS
COWBOY CAVE AND WALTERS CAVE

| | Edge Angles | | | |
| | Dart Points | | Arrowpoints | |
	Min.	Max.	Min.	Max.
Mean	38°	43°	35°	37°
Standard Dev.	7.1°	9.3°	5.6°	5.9°
No. of Specimens	100	100	24	24

| | Thicknesses (cm.) | |
	Dart Points	Arrowpoints
Mean	0.50	0.27
Standard Dev.	0.08	0.04
No. of Specimens	100	24

record at Cowboy Cave (see *Chipped Stone Projectile Points*).

Utilized Flakes

This was the largest single category of lithic artifacts recovered from Cowboy and Walters Caves. Utilized flakes were distinguished from debitage on the basis of various combinations of use retouch, wear polish, and minimal pressure retouch. Use retouch was the primary distinguishing characteristic, however; it was evident on 81 percent of the utilized flakes. Since use retouch is generally considered to be the result of cutting hard materials, the utilized flakes complement the knives, which exhibited use retouch on only 5 percent of the specimens. However, the mean minimum and mean maximum edge angles of 43° and 48°, respectively, for the utilized flakes were very similar to the knives.

Scrapers

In 71 percent of the scrapers recovered from Cowboy Cave, either the edge of a flake with an already existing steep edge angle was used, or the tool was pressure retouched unifacially to achieve a steep edge angle. The mean minimum and mean maximum edge angles were 66° and 69°, respectively, for all specimens.

Evidence of wear polish was found on 75 percent of the specimens, use retouch on 39 percent, and striations perpendicular to the edge on 29 percent. The scraper wear polish was both quantitatively and qualitatively different from the wear polish noted on the bifaces and utilized flakes. On many scrapers, a distinct rounding of the scraping edges was visible without magnification. In addition, the wear polish was not shiny, but rather dull and frosty in appearance.

Hammerstones

Fifty-seven hammerstones and hammerstone fragments were recovered from Cowboy and Walters Caves; 34 of them were unbroken. Out of the total of 57, 43 (75 percent) of the hammerstones were flaked by a hard hammer before being used for pounding. The mean weight of all unbroken specimens was 188 gm., although there was much variability in size, as is revealed by the standard deviation of 135 gm. All hammerstones were made of coarse-grained chert or quartzite.

Choppers

Six choppers were recovered from Cowboy Cave and three from Walters Cave. The chopping edges were formed by hard- and/or soft-hammer techniques, and exhibited mean edge angles of 67° (minimum) and 77° (maximum). The mean weight of the choppers was 156 gm.

Miscellaneous Tools

Burins, gravers, awls, drills, and small flaked grinders were recovered in small quantities (see Tables 7 and 8).

Cores

The cores from Cowboy and Walters Caves all exhibited hard-hammer or soft-hammer flake removal. Flakes, rather than blades, were the desired products. It should be noted that the two strata with the highest numbers of utilized flakes, Vb and IVc, also have the highest number of cores.

Debitage

The provenience data for the lithic debitage from Cowboy and Walters Caves are included in Tables 7 and 8. With the exception of Stratum IIb of Cowboy Cave (which is discussed below), the debitage from any particular stratum was of many different types of chert. The variability and generally small quanti-

ties of debitage suggests that there was no local source of chert, and also that very little knapping activity was performed in either cave. For most strata, the debitage was scattered in small quantities over the cave floor. Only three strata in Cowboy Cave, IIb, IVc, and Va, showed any localized concentrations of debitage.

Figure 20 shows the horizontal debitage provenience of Stratum IIb. The relatively high concentration of debitage in squares 14R26, 14R27, and particularly 15R27, can safely be interpreted as a knapping station, especially since the debitage was almost entirely a single type of gray blue chert. None of the other debitage concentrations in any stratum were homogeneous in chert type. In addition, 16 out of the 21 blanks and preforms recovered from Stratum IIb were from the same squares--14R26, 14R27, and 15R27. None of these blanks or preforms exhibited any wear polish.

Figure 21 (Stratum IVc) shows three localized concentrations of debitage, although light concentrations occur over most of the area.

Figure 22 (Stratum Va) shows two small areas of high debitage concentration at opposite ends of the cave, and a large, low concentration area towards the rear of the excavated area.

A probable explanation for the generally heterogeneous debitage (the single exception occurring in Stratum IIb), is that blanks of chert from various sources were carried to Cowboy Cave and reworked to form the desired tools as they were needed, although the many broken bifaces indicate that the reworking was not always successful.

Obsidian was conspicuously absent in the debitage from both caves. No obsidian tools were recovered, and only a single obsidian flake (from Test Trench 2) was found.

Lithic Heat Treatment

Evidence of heat treatment of chert was found in nearly every stratum of Cowboy Cave. The bifaces were the tool

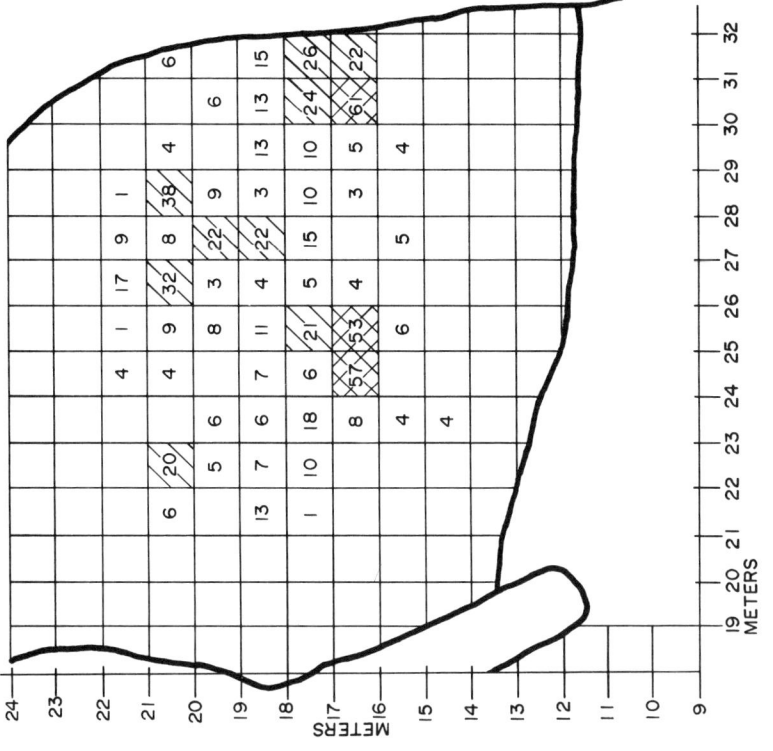

FIG. 21. Actual frequency of occurrence of debitage specimens by square, Stratum IVc. Densities are also indicated by intensity of hatching.

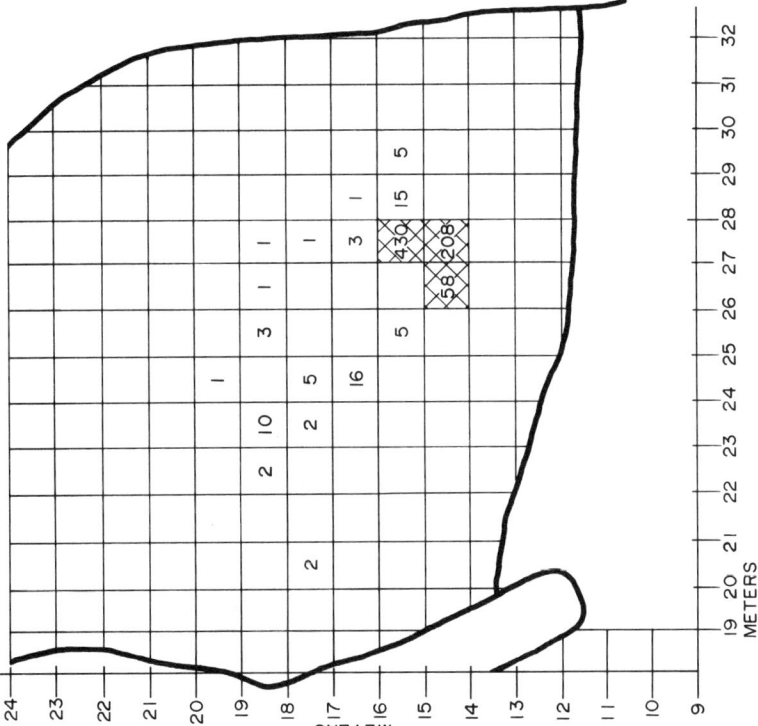

FIG. 20. Actual frequency of occurrence of debitage specimens by square, Stratum IIb. Densities are also indicated by intensity of hatching.

category which was most commonly heat treated. Typically, blanks and preforms were heat treated prior to further knapping. Since the identifiable characteristics of heat-treated chert are generally the result of the improper control of the heating and subsequent cooling processes, which result in broken or weakened chert (Purdy 1974 and 1975), the prevalence of successful heat treatment was difficult to recognize.

Crenated Cobbles
    Twenty fragmentary cobbles were recovered from Cowboy and Walters Caves.

All of these cobbles exhibited crenated fractures, which result from heating followed by rapid cooling (Purdy 1975). George Hyde (1959), in his ethnographic work among the Shoshoni, reports that heated rocks were dropped into water-tight baskets in order to boil food. Since dropping a heated cobble into water quickly cools the cobble, causing crenated fractures, these cobbles may be an indication of cooking activities.

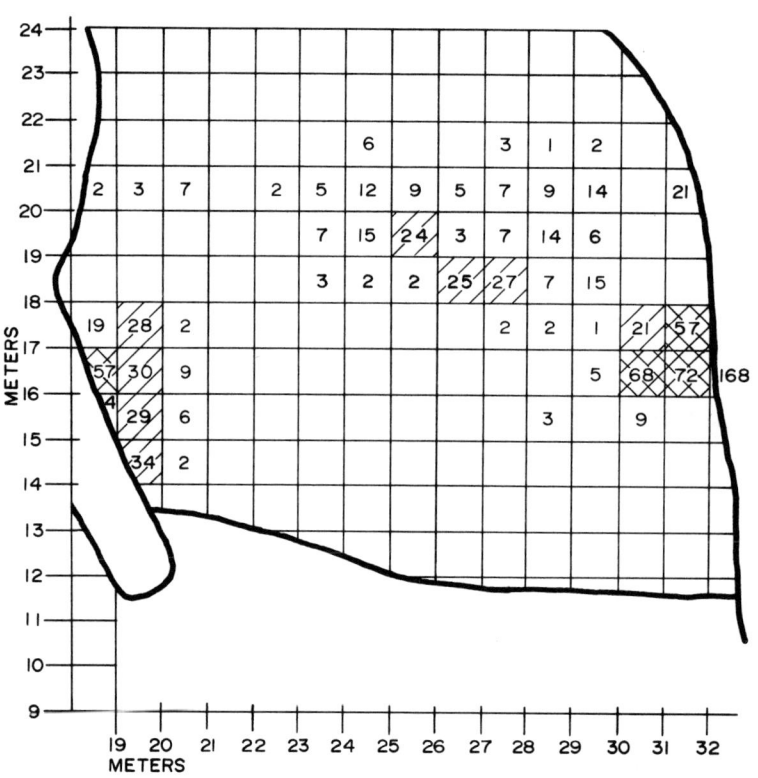

FIG. 22. Actual frequency of occurrence of debitage specimens by square, Stratum Va. Densities are also indicated by intensity of hatching.

FIBER ARTIFACTS

Nancy J. Hewitt

BASKETRY

Seventy-three pieces of basketry and various types of raw construction materials were recovered from Cowboy and Walters Caves. Since all pieces were fragmentary, no vessel shapes could be determined.

Considerable time was spent on positive identification of foundation elements, since they are the most commonly used diagnostic factors in basketry analysis. Adovasio (1970a:133) has shown that in the Great Basin there is a sequence of increasingly complex foundation patterns (starting with a single rod), that has some chronological significance. Examination of specimens was done with a hand-held glass; in some cases, dental picks were required for separating the elements. All specimens were coiled; they were divided into 12 subclasses using the following criteria: (1) Foundation--number and type of elements (rod, welt, bundle, etc.); (2) Type of stitch (interlocking, noninterlocking, split, intricate); (3) Work direction (right or left slant, or vertical); and (4) Tightness of stitch (open vs. closed). The subclasses established for basketry from Cowboy and Walters Caves are presented below. Proveniences for all basketry artifacts appear in Table 12.

Coiled Basketry

Subclass 1
Description: Open coiling, one-rod foundation, intricate stitch.
Number of specimens: 8.
Technique: See Figure 23a for foundation cross section.
Type of stitch: intricate (see Fig. 23h for diagram).
Work direction: left slant, eight.
Comments: There are actually two different kinds of intricate stitch represented in this subclass. Two of the specimens are similar to the type described from Danger Cave (Price 1957:245). The remaining six are similar to the type described from Hogup Cave (Adovasio 1970b: 140), the most common type in the Southwest according to Morris and Burgh (1941: 18).

Three specimens are the starts of basket bottoms, one is a wall fragment, one a rim fragment, and the rest are unidentifiable. One of the fragments of the Danger Cave type appears to be pitched, which seems useless if done for water tightness, since the stitches are spaced so far apart. The largest and best-preserved specimen (Fig. 24a) demonstrates how the vessel was shaped by adding or dropping stitches. Tightness of stitch varies from one to two stitches per centimeter.
Comparisons: Open coiled baskets of this subclass have been found at Danger

TABLE 12

DISTRIBUTION OF BASKETRY, SANDALS, AND WORKED PLANT FIBERS FROM COWBOY CAVE

	NP	TT1	II a	II b	III a	III b	III c	III d	III e	III f	III g	III h	III i	IV a	IV b	IV c	IV d	V a	V b	V c	V d	Sur	TOTAL
Coiled Basketry																							
Subclass 1	2						2	1						2					1				8
Subclass 2						1										1		1					3
Subclass 3	3										1		1	1		1	3	2	1	1			14
Subclass 4	2			1			1					1		1		2	1	1	2				11
Subclass 5	1						2					3		1		2	1		5	1			14
Subclass 6							2							1		1	1	5		1		1	8
Subclass 7	1																						1
Subclass 8	1						2							1					1				5
Subclass 9																				1			1
Subclass 10										1					1			1					2
Subclass 11										1					1						1		2
Subclass 12							1																1
Problematical Objects							1					1				1						1	11
Yucca Strips	60		6	1		2	20	6	65	16	16	27	3	3		5	5	11	4	5		5	251
Sandals																							
Plain Weave	6				1	2	1	2		9	9	2		1	2	3			2	1			40
Open Twined	4		2			1	1		1	1						1	1					1	13
Cross Weave																1							1
Worked Plant Fibers																							
Grass and Sagebrush Bark Pads							2	2				1		2		2	7		1				17
Fiber Bundles	3						1	1				2	1	1	1	1	1	3					16
Coils	6	1	2							1		2		2		6	3		2	1		2	25
Wrapped Rings	2															1	1	2					6
Elongated Split Rings	4											4		1	1	1	1		3	1	1		18
Buttons														1		6		3					9
Bent Twigs and Splints																3	2	1					5
TOTAL	94	1	11	1	3	24	10	70	28	35	34	7	2	44	26	17		47	7	9		12	482

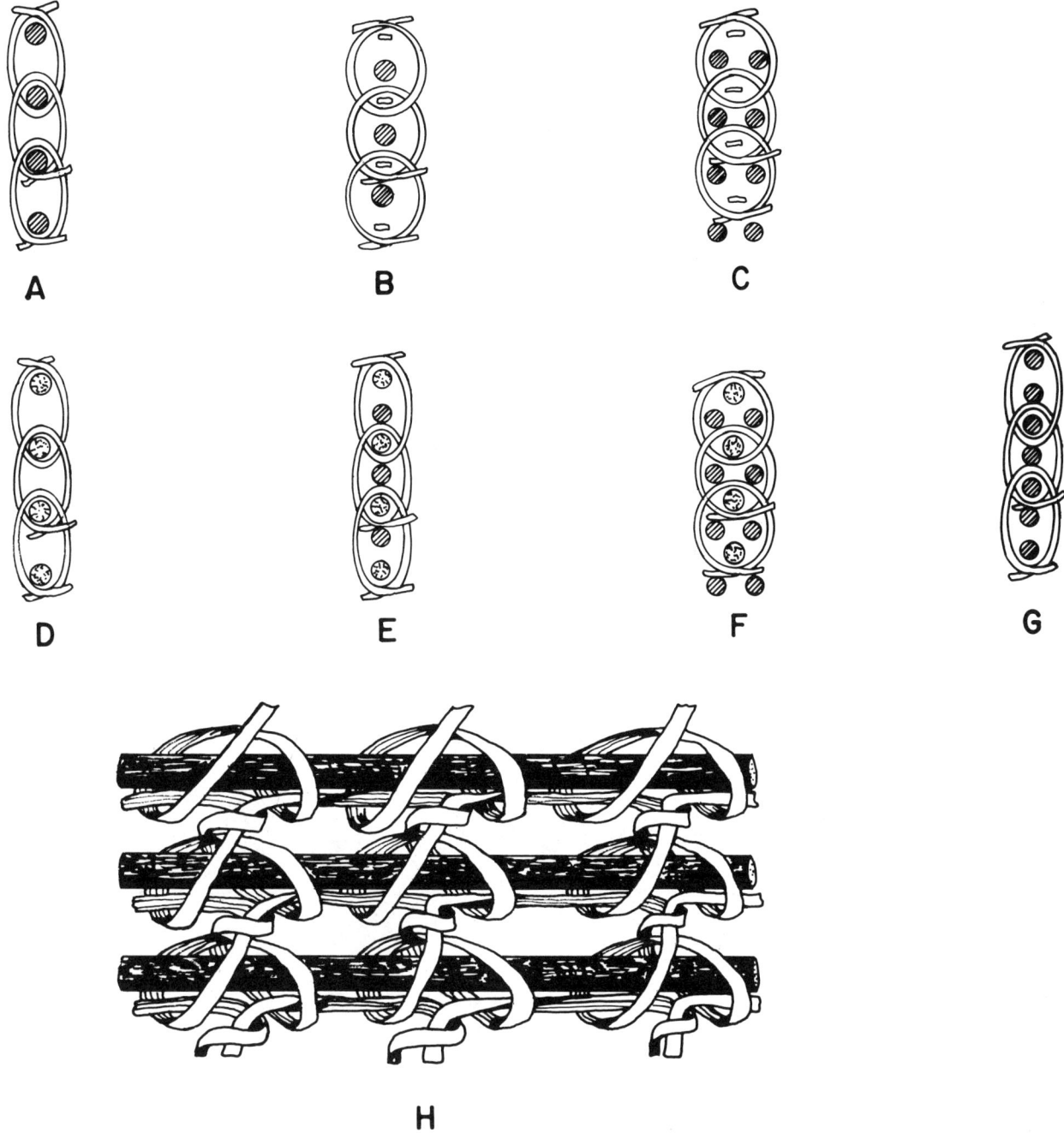

FIG. 23. Cross sections of coiled basketry foundations: a, one-rod; b, one-rod and welt stacked; c, two-rod and welt; d, bundle; e, one-rod and bundle stacked; f, two-rod and bundle; g, two-rod stacked; h, open coiling, intricate interlocking stitch (reproduced from Adovasio 1970).

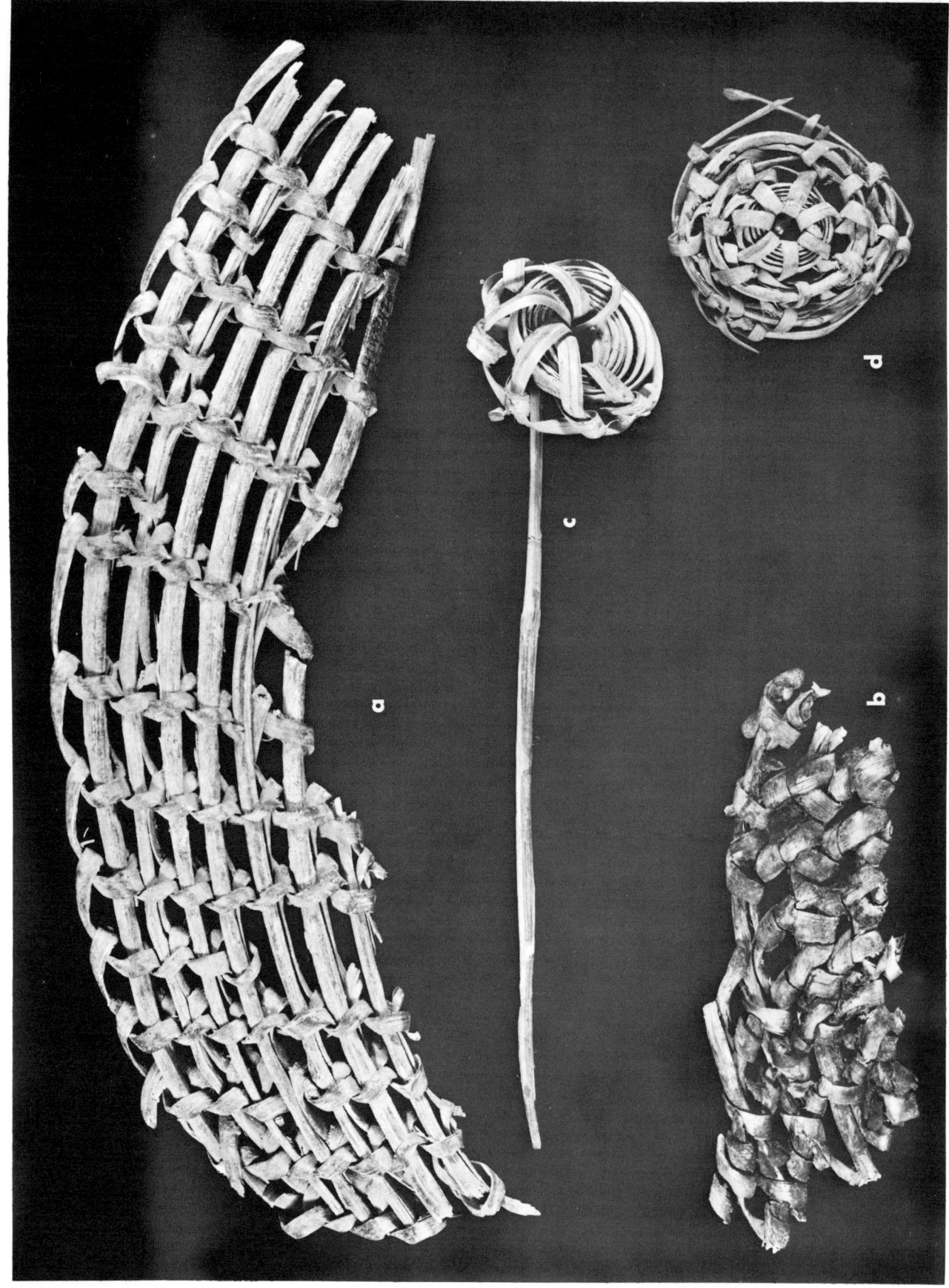

FIG. 24. Basketry. Examples of spaced coiling with intricate stitch. Note that c and d are starts of baskets; d is 3 cm. in diameter.

Cave (Price 1957:245) dating from 4900 to 4000 B.P.; Hogup Cave (Adovasio 1970b:140) dating 6190 B.P.; DuPont Cave (Nusbaum 1922:96-97); Etna Cave (Fowler 1973:24); Promontory (Steward 1937:34); and in the Anasazi region at Lake Canyon (Weltfish 1932:19); Prayer Rock District, and East Canyon (Morris and Burgh 1941: 17). Morris and Burgh claim that this is a distinctive Basketmaker trait (1941:18). Adovasio (1970a:78) notes the occurrence of this technique in Utah, Idaho, and Wyoming, but claims there is no history of its development in the Basin, and the pieces might be tradeware. However, three of the specimens from Cowboy Cave are the starts of coiled baskets, evidence that the inhabitants of the cave were familiar with the technique and employed it in their basketry construction. Thus, it would appear that open coiling with an intricate stitch was an indigenous technique at Cowboy Cave, and that these specimens do not represent tradeware.

Subclass 2
   Description: Close coiling, one-rod foundation, interlocking stitch.
   Number of specimens: 3 (see Table 12).
   Technique: See Figure 23a for foundation cross section.
   Work direction: right slant, one; vertical, one; unidentifiable, one.
   Comments: Two of the three specimens are the starts of basket bottoms. One is very fine and fragile; the stitch cannot be definitely determined. The third specimen is fragmentary. Its stitch is consistently two stitches per centimeter, with none of the stitches split.
   Comparisons: Coiled baskets of this subclass are known from Danger Cave from levels D V (4900 to 4000 B.P.), D IV (2819 B.P.), D III (4950 B.P.) (Price 1957:245); Ventana Cave, dated at 950 to 550 B.P. (Haury 1950:405); Grand Gulch, Utah (Weltfish 1932:10); Battle Canyon (Weltfish 1932:15); Coombs Village (Lister et al. 1960:261); Buffalo Cave (Lambert 1961:64); Humboldt Cave (Heizer and Krieger 1956:46); Evans Mound and

Caldwell Village (Ambler 1966:245); and Yampa Canyon (Burgh and Scoggin 1948:56). Rudy (1953:154) reports a few samples, Morss (1931:74) reports this technique from the Fremont River region, and Morris and Burgh (1941:7) find this class in the Basketmaker III and Pueblo I and III periods of the Southwest. Most importantly, this technique is prevalent at Sand Dune Cave (Lindsay et al. 1968:120) in the Desha Complex levels dating between 7450 and 7950 B.P. It is Adovasio's hypothesis (1970a:152) that this technique is the oldest class of coiling in the Basin, and that all other coiling techniques are elaborations of the basic one rod foundation. One of the specimens from Cowboy Cave does have an early date of 6675± B.P. However, this is not the earliest coiling technique at Cowboy Cave (see Table 12).

Subclass 3
   Description: Close coiling, one-rod and welt, noninterlocking stitch.
   Number of specimens: 14.
   Technique: See Figure 23b for foundation cross section.
   Type of stitch: noninterlocking.
   Work direction: left slant, two; right slant, seven; vertical, three; non-identifiable, three.
   Comments: This subclass is one of several one-rod variations, with a secondary foundation, a welt, added. By definition (Adovasio 1970a:18), a welt is "an element used in coiled foundations in conjunction with one or more rods. The welt is generally a small, flattened twig which is stacked vertically on a single rod or used in a triangular arrangement with two or more rods."
   Subclass 3 includes three wall fragments, five base fragments, and six fragments of undetermined form. None of the fragments are caulked; five have been charred. Most of the specimens display a tightness of stitch of three stitches per centimeter. Three pieces have a few sporadic split stitches.
   Comparisons: This subclass, as defined here with a noninterlocking stitch, has not been recorded previously in any available literature. Therefore, this

technique is deemed to be unique to Cowboy Cave.

Subclass 4
  Description: Close coiling, one-rod and welt, split stitch.
  Number of specimens: 11.
  Technique: See Figure 23b for cross section of foundation.
  Type of stitch: split.
  Work direction: left slant, two; right slant, seven; vertical, two.
  Comments: Essentially Subclass 4 is identical to Subclass 3 except for the difference in stitch type. In many textile reports it appears that both subclasses are combined. If the specimens from Cowboy Cave were given the same classification, the resulting subclass would be the most popular technique, with 25 specimens representing a third of the total collection. However, the stitch types were separated in this analysis to see if there were any chronological preference for one over the other. The provenience chart (Table 12) shows that, except for one instance, both types seem to share popularity through the same time span.
  Only one piece from this subclass can be identified as a wall fragment; the remaining specimens are too small to determine form. Six fragments are charred. Closeness of stitch varies from two to four stitches per centimeter.
  Comparisons: Specimens of this subclass have been reported from Danger Cave (Price 1957), Ventana Cave (Haury 1950:406), Lovelock (Loud and Harrington 1929:672), Hogup Cave (Adovasio 1970b: 141), Oregon Caves (Cressman 1942) and Humboldt Cave (Heizer and Krieger 1956: 46). Due to the similarity of technique, it is possible that this subclass is grouped with Subclass 6 in many of the reports and may actually be more widespread in occurrence. Historically, Mason (1902:502) reports this technique among the Zuni and Utes.

Subclass 5
  Description: Close coiling, one-rod and bundle, noninterlocking stitch.

Number of specimens: 14.
  Technique: See Figure 23e for cross section of foundation.
  Type of stitch: noninterlocking.
  Work direction: right slant, ten; left slant, two; unidentifiable, two.
  Comments: Basically, this technique is similar to Subclass 3 except the secondary foundation is a bundle of fibers, as opposed to a welt. The use of the bundle in coiled basketry renders the vessels virtually watertight when wet, making pitching or caulking unnecessary.
  Six specimens from this subclass are large enough to be identified as wall fragments, three of which have mending stitches. One of the pieces was also mended with a piece of Z-twist cordage. One fragment appears to be part of a rim. Some of the stitches on one specimen appear to be accidentally split because their occurrence is irregular and without patterning.
  Comparisons: Basketry displaying this technique has been reported from Promontory Caves (Steward 1937:118), Hogup Cave (Aikens 1970:141), Median Village (Marwitt 1971:54), Uncompahgre Plateau (Wormington and Lister 1956:32), Danger Cave (Price 1957:247), Old Woman (Taylor 1957:46-47), the Anasazi area (Morris and Burgh 1941: 68), Dinosaur National Monument (Burgh and Scoggin 1948:56), Caldwell Village (Ambler 1966:65), the Colorado River area (Judd 1926:148), and the Fremont River area (Morss 1931:73). Apparently this technique is not known historically.

Subclass 6
  Description: Close coiling, one-rod and bundle, split stitch.
  Number of specimens: 8.
  Technique: See Figure 23e for foundation cross section.
  Type of stitch: split.
  Work direction: right slant, four; vertical, three; nonidentifiable, one.
  Comments: Three specimens from this subclass are the bottoms of baskets, one apparently from a flat basket. Frayed cordage fragments protrude through the stitches of two of these specimens. One other piece can be identified as a wall fragment; the rest are too small to

determine form. The closeness of stitch is predominantly three stitches per centimeter. It should be noted that this subclass is virtually the same as Subclass 5 except for the type of stitch. Table 12 shows that the split stitch appears later along the time scale. When the subclasses are combined, the technique of one-rod and bundle represents the second most popular technique at Cowboy Cave. When compared with the predominant technique of one-rod and welt, it can be seen that both coexisted through time.

Comparisons: Subclass 6 textiles are reported from Danger Cave (Price 1957:247), Hogup Cave (Adovasio 1970b: 141), Promontory Cave (Steward 1937:34), Fremont River area (Morss 1931:73), Etna Cave (Mulloy 1958:118).

Subclass 7
Description: Close coiling, two-rods stacked, interlocking stitch.
Number of specimens: 1.
Technique: See Figure 23g for cross section of foundation.
Type of stitch: interlocking.
Work direction: right slant, one.
Comments: The single specimen of this subclass is too small to determine form. Some of the stitches are "double", thereby passing over more than three rods. The purpose of these stitches is unknown.
Comparisons: This subclass is quite rare throughout the Basin, the Southwest, Nevada, and even in the Oregon area (Adovasio 1970a), occurring in only two other sites. Haury reports finding one fragment at Ventana Cave and doubts its archeological occurrence elsewhere (1950:407). One other specimen (according to Adavasio 1970a:30) was found at Chimney Cave in Oregon.
Ethnohistorically, the Utes are reported to be the only group who utilized this technique (Weltfish 1930:472).

Subclass 8
Description: Close coiling, two-rod stacked, split stitch.
Number of specimens: 5.
Technique: See Figure 23g for foun-

dation cross section.
Type of stitch: split.
Work direction: right slant, three; vertical, one.
Comments: The form of the five specimens of this subclass could not be determined. Two pieces appear to be caulked with pitch; one piece was charred. Closeness of stitch varied from two to three stitches per centimeter.
Comparisons: An examination of available literature shows this technique as described has not been reported at any other site in the western states.

Subclass 9
Description: Close coiling, two-rods and welt, interlocking stitch.
Number of specimens: 1.
Technique: See Figure 23c for foundation cross section.
Type of stitch: interlocking.
Work direction: vertical, one.
Comments: The single specimen of this subclass is a charred fragment with four to five stitches per centimeter. As shown in the cross section in Figure 23, the foundation of this subclass consists of two horizontal rods surmounted by a welt.
Comparisons: Although the two-rod and welt technique is not reported by Price from Danger Cave (1957), a reanalysis of the material by Adovasio (1970b: 146) shows that this subclass was combined with the three-rod bunched subclass, but has a noninterlocking stitch. This technique has also been reported at Hogup Cave (Adovasio 1970b:141), Lovelock Cave (Loud and Harrington 1929:76), and Humboldt Cave (Heizer and Krieger 1956:46), but all of these specimens have a split, rather than an interlocking, stitch.

Subclass 10
Description: Close coiling, two-rod and bundle, noninterlocking stitch.
Number of specimens: 2.
Technique: See Figure 23f for foundation cross section.
Type of stitch: noninterlocking.
Work direction: vertical, one; non-identifiable, one.
Comments: Both specimens are fragmentary and in poor condition. One piece

is charred. The tightness of the stitch can be determined on only one specimen; it is five stitches per centimeter.

Comparisons: Typically, this technique is characteristic of Basketmaker II (and later) basketry. Sites in the Southwest where this technique is reported are too numerous to list here. However, we should note that the technique has been reported at Sand Dune Cave in the so-called mixed Basketmaker/ Desha levels (Lindsay et al. 1968:97), and at Ventana Cave (Haury 1950:406). Apparently, this technique has not been reported from other sites in the Basin, and its occurrence at Cowboy Cave represents diffusion of technique or tradeware from the Southwest.

Subclass 11
Description: Close coiling, bundle foundation, noninterlocking stitch.
Number of specimens: 1.
Technique: See Figure 23d for foundation cross section.
Type of stitch: noninterlocking.
Work direction: left slant, one.
Comments: This subclass is represented by one specimen which is too fragmentary to determine form. The stitch is fairly tight, from two and a half to three stitches per centimeter. Actually, the foundation of this specimen is not a bundle in the true sense of the term, in that it is not a bunch of fibrous material, but rather an combination of fibrous material and small rods.
Comparisons: This technique, utilizing a bundle as a foundation, was apparently rare in the Basin, and occurred fairly late in the Southwest. Two specimens with an interlocking stitch were reported from Caldwell Village dating ca. 1430±70 B.P. (Ambler 1966:65). The same technique, again with an interlocking stitch, was present at Uncompahgre Plateau (Wormington and Lister 1956:121), as well as from Sand Dune Cave (Lindsay et al. 1968:97). Morris and Burgh (1941:10) claim that the bundle technique with a noninterlocking stitch was common in Basketmaker III and Pueblo IV times. Because

of its apparent late occurrence in the Southwest, Adovasio (1970a:87) suggests that the technique represents diffusion from Mexico.

Modern Pima and Papago utilize this technique almost exclusively in their basketry construction (Haury 1950:403).

Subclass 12
Description: Close coiling, splint foundation, noninterlocking stitch.
Number of specimens: 1.
Technique: single splint foundation.
Type of stitch: split.
Work direction: right slant, one.
Comments: This solitary specimen is the beginning of a basket bottom. Tightness of stitch could not be determined. Although it is not certain that this technique constitutes a subclass of its own, it is different enough from the other subclasses that it could not be included with one of them.
Comparisons: The only other report of this technique was in Dinosaur National Monument by Burgh and Scoggin (1948:58). However, that specimen utilized an interlocking stitch.

Unidentifiable Basket Fragment
The slightly conical bottom of a coiled basket is so heavily pitched that the construction technique cannot be determined. It measures approximately 5 cm. by 5 cm. There are several mending stitches. This is one of the few pitched pieces of basketry from Cowboy Cave and comes from Stratum Va.

Basketry Material

Five pieces of the total basketry collection were examined to determine which plants were utilized. All pieces identified, both rods and stitching elements, were *Rhus trilobata.*

Basketry Construction Components

BASKETRY WITHES. Four large bundles of basketry withes were probably prepared for use as stitching elements or welts.

BASKETRY ROD. Found with two of the basketry withe bundles was a single stick, looped by wrapping one end around the other. This stick may have been intended for use as a basketry rod, and was looped in order to be worked easily into a coiled basket.

GRASS BUNDLES. Two bundles of unprepared grass, with roots still intact, may be raw materials for the bundles used in basketry construction. One is loosely tied around the top, the other is merely folded over.

Summary and Conclusions

Although 12 different basketry techniques are represented in the Cowboy Cave collection, two techniques overwhelmingly outnumber the rest. The most popular technique is one-rod and welt, represented by Subclasses 3 and 4, and totaling 25 specimens, or about a third of the total collection. The second most popular technique is one-rod and bundle, represented by Subclasses 5 and 6, and totaling 22 specimens. The similarities between the two techniques are obvious; in fact, many reports combine them under one technique. The provenience chart, Table 12, shows that both techniques enjoyed popularity simultaneously, although two specimens of the one-rod and welt technique appear fairly early. Adovasio (1970a:50) has shown that the one-rod and welt technique appears first in the Basin in Archaic times, then is gradually replaced by the one-rod and bundle technique which is characteristic of the Fremont culture. The occurrence of the latter technique may, in fact, represent the transition to the Fremont culture. At this point it should be noted that one-rod (Subclass 2); one-rod and welt, split stitch (Subclass 4); and one-rod and bundle (Subclasses 5 and 6); are techniques that are present fairly early in the Basin and continue into one of the more recent prehistoric cultures in the Basin, which is essentially Fremont (one-rod and bundle being predominant).

Although the specimens are few, a third important basketry technique represented at Cowboy Cave is one-rod with an interlocking stitch (Subclass 2). This technique is significant because it can be tied in with the Desha C Complex as outlined by Lindsay et al. (1968:120). The technique occurs elsewhere in the Basin and western Nevada as well.

Another important basketry technique found at Cowboy Cave was spaced coiling with an intricate stitch (Subclass 1). As mentioned above, Morris and Burgh (1941:18) see this technique as a distinctive Basketmaker trait. Further, Adovasio (1970a:78) claims the occurrence of this technique in Utah, Idaho, and Wyoming represents the limits of Southwestern influence, but considers it a tradeware, since there is no history of its development in the Basin. Some of the specimens from Cowboy Cave, however, are the unfinished starts of basket bottoms, which indicates that the inhabitants of the cave were familiar with the technique and utilized it in their basketry construction. The purpose of these kinds of baskets is not known; the spaced coiling is too open for winnowing. Aesthetic value cannot be ruled out, and may in fact be the motive for the development of the technique. It appears, then, that this technique, as represented at Cowboy Cave, is not tradeware but either developed indigenously, or was a trait acquired from the Southwest and used locally.

Since the occurrence of two-rod techniques collectively is only 11 percent, their significance at Cowboy Cave is not clear. Only two specimens of the typical two-rod and bundle Basketmaker technique occur in the upper levels of Unit V. The two-rod stacked specimens found here are rare or nonexistent elsewhere, as is the single two-rod and welt specimen. It should be noted that the two-rod and bundle Basketmaker technique occurs later than the so-called Basketmaker technique of spaced coiling with an intricate stitch.

The early occurrence of the bundle with a noninterlocking stitch (Subclass 11) is questionable. Morris and Burgh (1941:10) claim that this technique was used in the Southwest in Basketmaker III and Pueblo IV times. However, as mentioned above, the foundation in the specimen is not a bundle in the pure sense of the term and may represent a unique single instance, or a technique that has not been previously described.

SANDALS

Seven whole and 47 fragmentary sandals were recovered from Cowboy Cave. Two fragments were found in Walters Cave. These sandal remains represent three distinct sandal types: open-twined, plain-weave, and cross-weave.

Open-twined Sandals

One complete sandal and 12 sandal fragments of this type were recovered. All are made of whole yucca leaves. Typically, two or more leaves are treated as one element to form the warps. The average width of these warps is 1 cm.; each of them is doubled back at the toe end and held in place by a row of Z-twist yucca twining. Usually there are 10 warp elements. Five to six additional rows of twining are spaced along the length of the sandal, creating a very open weave. The distance of spacing varies from 2.5 cm. to 6 cm., the closer weave being toward the ends. Several of the specimens contain inner-soles of grass padding. Only two specimens were complete enough to measure length and width; one is 24 by 12 cm., the other is 22 by 10 cm. All specimens show considerable wear; most of the elements are frayed, and, in many cases, the entire heel section of the sandal is worn through. The shape seems to be roughly rectangular, with a rounded toe. Many knotted yucca strips attached to, or found with, the sandals may be remnants of ties, but give no clue as to how the sandals were attached to the foot. (See Fig. 25, and Fig. 26a, b.)

Specimens identical to these open-twined sandals were recovered from Sand Dune and Dust Devil Caves, and form part of the basis for the definition of the Desha Complex (Lindsay et al. 1968:95).

What appears to be part of an open-twined sandal from Bernheimer Alcove is illustrated in Sharrock, Day, and Dibble (1963:207, Fig. 776). It is classified, however, as a twined mat or container. Hurst (1947, Pl. I, Fig. 3, Pl. II, Fig. 9) reports similar sandal specimens from Dolores Cave. He believed the site to be no earlier than Basketmaker II, whereas the Cowboy Cave evidence is that it is probably much earlier.

Open-twined sandals from Cowboy Cave first appear in Stratum IIb, which dates to 8575±80 B.P. The rest appear sporadically in Unit III, dating between 7215±75 B.P. and 6675±75 B.P. (dates comparable to dated Desha Complex specimens), late in Unit IV (3635±55 B.P. to 3330±80 B.P.), and early in Unit V (1890±65 B.P.). Specific proveniences appear in Table 12.

Plain-weave Sandals

More than 70 percent of the total sandal collection has been classified as plain weave. This type employs a basic over-one-under-one weaving technique. The warp elements consist of two to four yucca leaves treated as one. They appear to be a continuous element, though they may be spliced within the weave. The weft elements, consisting of two to four leaves, are continuous, and double back into the weave at the edge of the sandal (see Figs. 26c and 27). These elements vary from one to two centimeters in width. Both warps and wefts are evenly spaced between two and five mm. apart.

Five specimens of this sandal type are whole; 35 are fragmentary. The average length of whole specimens is 19 cm.; the average width is 11 cm.

The method of tying the sandals to the feet was relatively simple. Thin yucca strips were looped in several places around the warp at each edge of the sandal. These strips from both sides would cross the foot, meeting in the middle. Here they were either looped or were knotted together, using a variety of knots (square, granny half-hitch, sheetbend, overhand slip). For a diagrammatic sketch of a sandal tie, see Fig. 28. One sandal has a single yucca strip across the toe which is secured to the warp on each side.

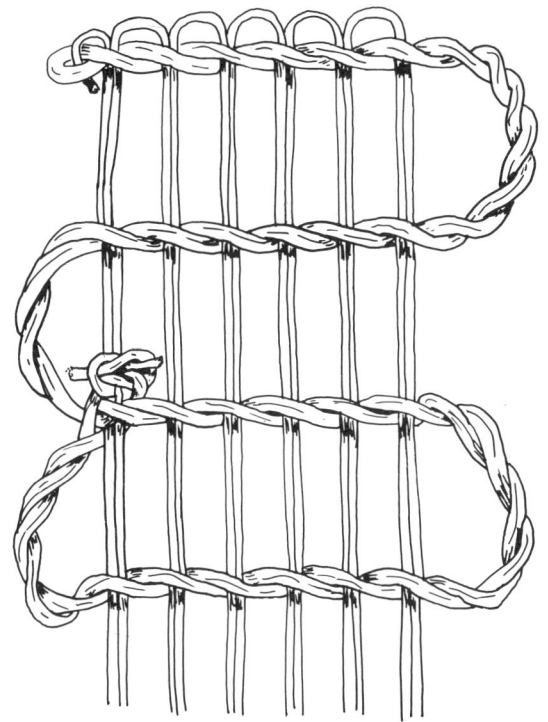

FIG. 25. Diagrammatic view of open-twined sandal (after Lindsay et al. 1968).

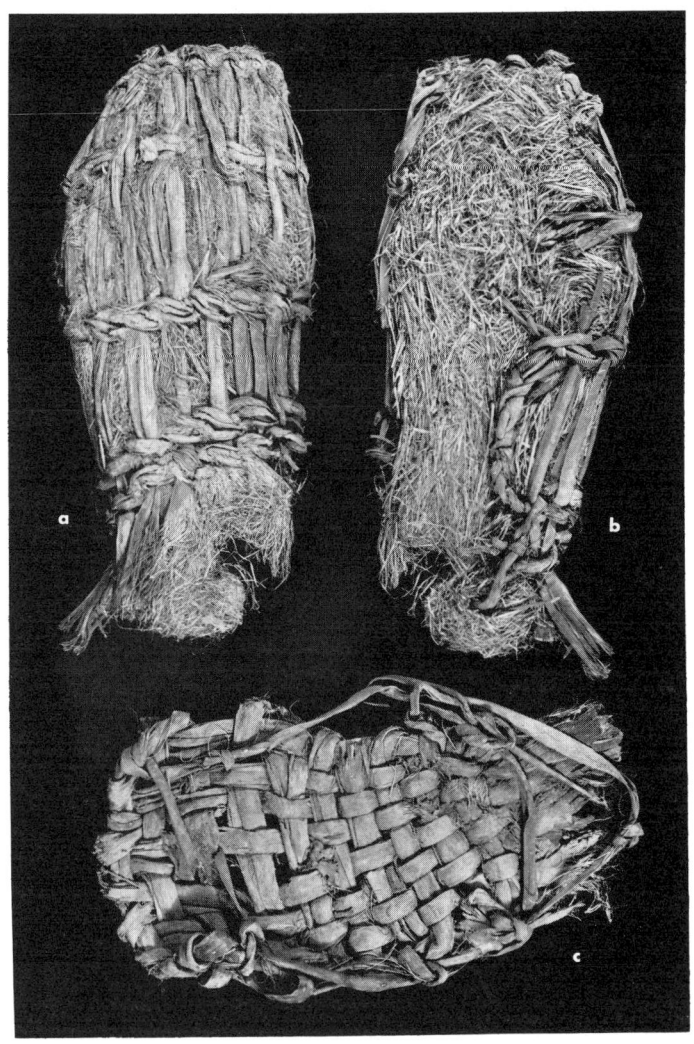

FIG. 26. Sandals: a, bottom view of open-twined sandal (24 cm. in length); b, top view of same open-twined sandal with grass pad; c, plain-weave sandal (20 cm.).

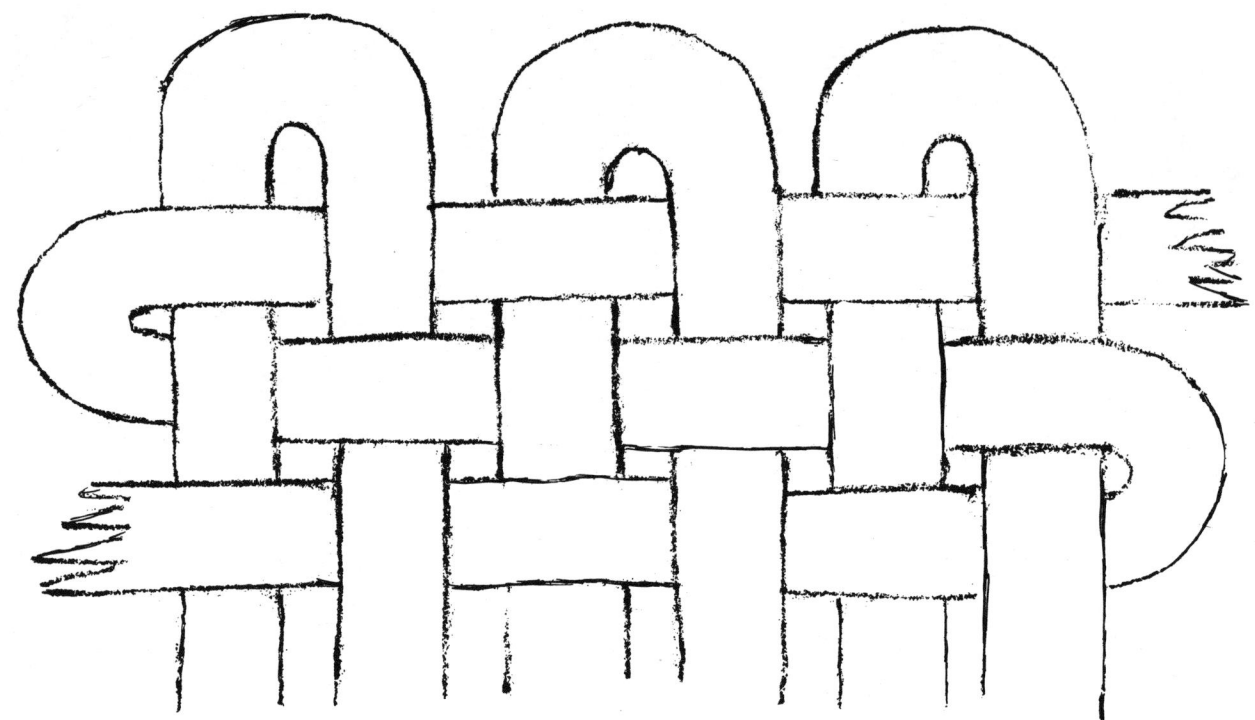

FIG. 27. Diagrammatic view of plain-weave sandal.

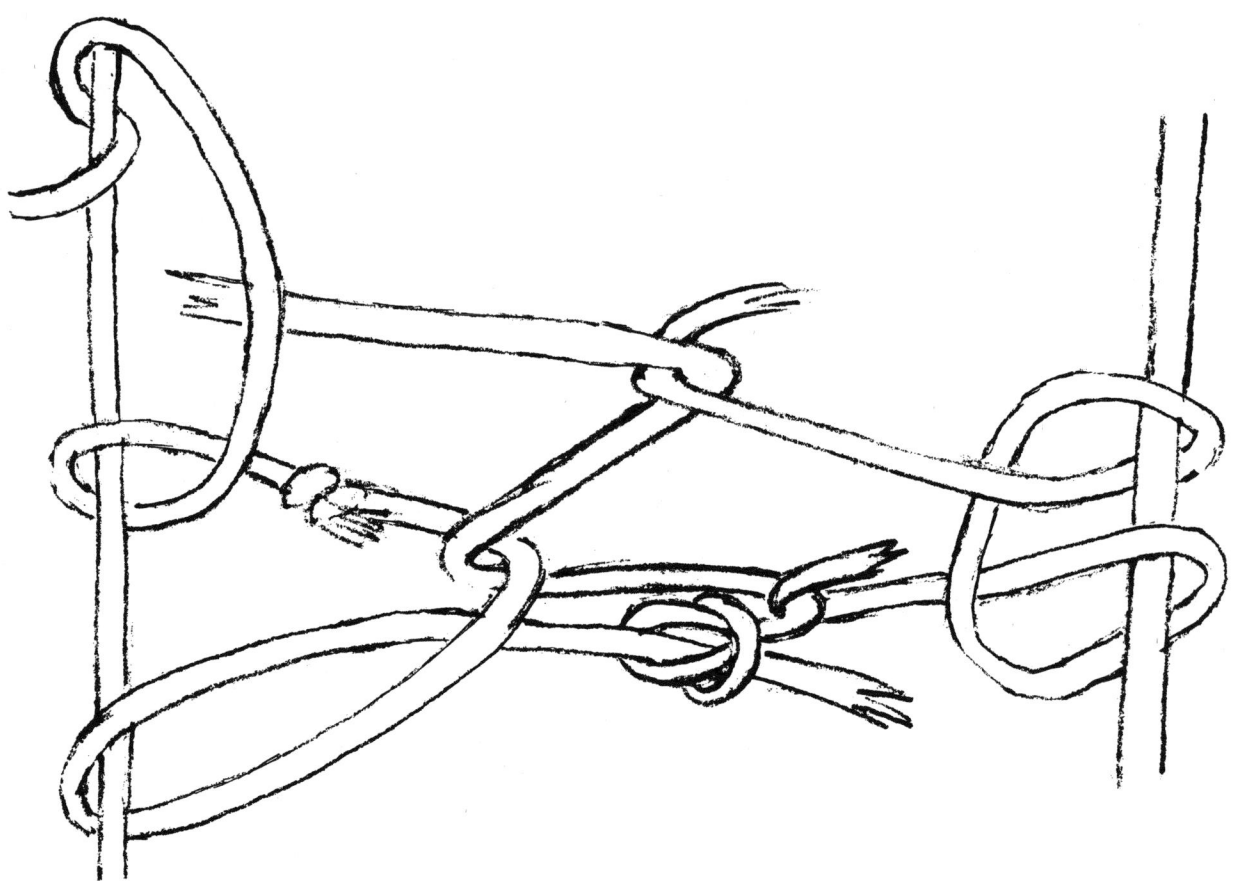

FIG. 28. Diagrammatic view of tie on plain-weave sandal.

Occasionally, pieces of cord were used as ties.

Yucca is the construction material used for all but one of the sandals. The exception is made of bark--either juniper or sagebrush. Eight sandals retain grass padding innersoles. One small fragment from a plain-weave sandal has a single row of twining at one end. The sandals are roughly rectangular with rounded ends and are crude in general appearance.

Plain-weave sandals occur almost continuously through Units III and IV, with one occurring late in Unit V, and two recovered from the surface (see Table 12).

According to available literature, this sandal type as named does not occur elsewhere. However, an illustrated sandal designated "Coarse Warp Face" by Lindsay et al. (1968:118, Fig. 94d) is identical to our plain-weave type. Stratigraphically this type is identified as part of the Desha Complex.

Cross-weave Sandal

This sandal type is represented by a single specimen from Cowboy Cave. Both warp and weft elements are two-ply-S-twisted yucca fiber, approximately 5 cm. wide. There are six warps interwoven by the weft elements in a simple over-one-under-one pattern. However, the ends of each weft element are left out of the weave at the middle of the sandal creating a fibrous pad (see Fig. 29a, b). It is difficult to determine whether this pad is the bottom or the top of the sandal. The weaving is very tight and gives the superficial appearance of twining. This specimen, nearly whole, measures 14 by 8 cm. and has a rounded toe (the heel portion is missing).

Kidder and Guernsey (1919:158, Pl. 67) have classified this technique as type 1, b, which characterizes the Basketmakers. The specimen from Cowboy Cave was found on the surface.

CORDAGE

There were 483 lengths of cordage made from plant fibers recovered from Cowboy Cave, and 18 from Walters Cave. Analysis of the collection included determining the direction of twist, the diameter of cordage, the material employed, manufacturing techniques, and knots used.

Relatively few artifacts made of cordage were recovered. These include 12 net fragments, an apron or string skirt, two necklace fragments, three unclassified objects, and what seem to be fragments of twined grass mats.

The distribution of cordage, fibers, and knots appears in Tables 13 and 14. Most of the cordage is small scraps which show the effect of hard and continual usage. The average length is 8 cm. Some are burned, worn to the point of unraveling, have broken fibers, or are even worn through. Many scraps are frugally knotted together in an effort to save cordage. In short, no one wasted string.

DIRECTION OF TWIST. In a description of cordage, the term *ply* refers to a single yarn which is usually plied with another single yarn to become a two-ply cord or yarn. The direction of twist is determined as follows: "If the elements are twisted in one direction so that the slope of the spirals, when held in a vertical position, conforms to the central portion of the letter S, the cord is said to have an S-twist. If the elements are twisted in the opposite direction, the cord has a Z-twist." (Rohn 1971:114). Two-ply cords are by far the most common, although other plies occur and will be discussed later.

A tabulation of the cordage specimens from Cowboy Cave according to the direction of twist shows 233 pieces of Z-twist cordage, compared with 204 pieces of S-twist cordage. Both twist types are present throughout all strata. Z-twist is predominant in Units II and III, but above Unit III, S-twist gains a slight predominance.

There are several methods of spinning yarn with a spindle; one of the most common is to roll it along the thigh. Ruth Underhill (1944:36) has shown that twist direction is dependent upon the direction the spindle is rolled. If it is rolled away from the body, an S-twist cordage

TABLE 13

DISTRIBUTION OF CORDAGE, CORDAGE MATERIALS, AND KNOTS USED IN CORDAGE, COWBOY CAVE

Unit:		II		III											IV				V					
Stratum:	NP	a	b	a	b	c	d	e	f	g	h	i	j	k	a	b	c	d	a	b	c	d	Sur	TOTAL
Cordage Types																								
One Ply	6						2	1	5	3	1	3			4	1	1	3	1	4		2	2	39
Two Ply Z-twist	44	4	1			1	5	6	34	19	16	16		2	22	5	14	5	5	16	2		16	233
Two Ply S-twist	41	2	5				4	4	15	12	12	9		1	13	2	21	16	7	20	4	3	13	204
Two Ply Hawser (SS)																1								1
Multitwist Semi-Hawser	1														1				2	1				5
Multitwist Cable																1								1
Three Strand Braid									2								1							3
Ten Strand Braid	1																			1				2
Doubled Back	5		1						2	1	2				1	1		1					1	15
Netting Fragments	2								5	1	1							1		1			1	12
Cordage Materials																								
Apocynum-Asclepias	57	2	4			1	5	9	28	18	19	12		3	17	4	26	18	12	26	5	3	14	283
Grass	20	1	2				3	2	19	6	6	9			8	1	8	8	4	9	1		1	108
Yucca	13								7	6	1				13		2			2				44
Artemisia			2				1			2		6			2	3				1				17
Linum																							13	13
Juniperus	1														1							1	1	4

Material								Total
Cowania					1			1
Gossypium	1							1
Hair			1			1	1	2
Combined Fibers								
Grass and Yucca	1	3	2	1	2			10
Grass and *Apocynum-Asclepias*		1	1		1			3
Grass and *Juniperus*			1					1
Apocynum-Asclepias and Yucca	1		1			2		3
Apocynum-Asclepias Yucca, and Grass		1	1					2
Apocynum-Asclepias Tied to Fur	1			1				2
Knots								
Overhand	8	9	3	3	2	3	10	48
Square	9	8	4	1	3	1	1	40
Sheetbend	6	2	1	1	1	1	1	13
Lark's-head	1	13	2	1	1	1	1	19
Granny	3							3
Slip	1		1	1				3
Non-identifiable	5	2	4	1	3	3	1	31

	Sheetbend	Lark's-head	Square	Overhand	Slip
Netting Knots	106	28	4	4	1

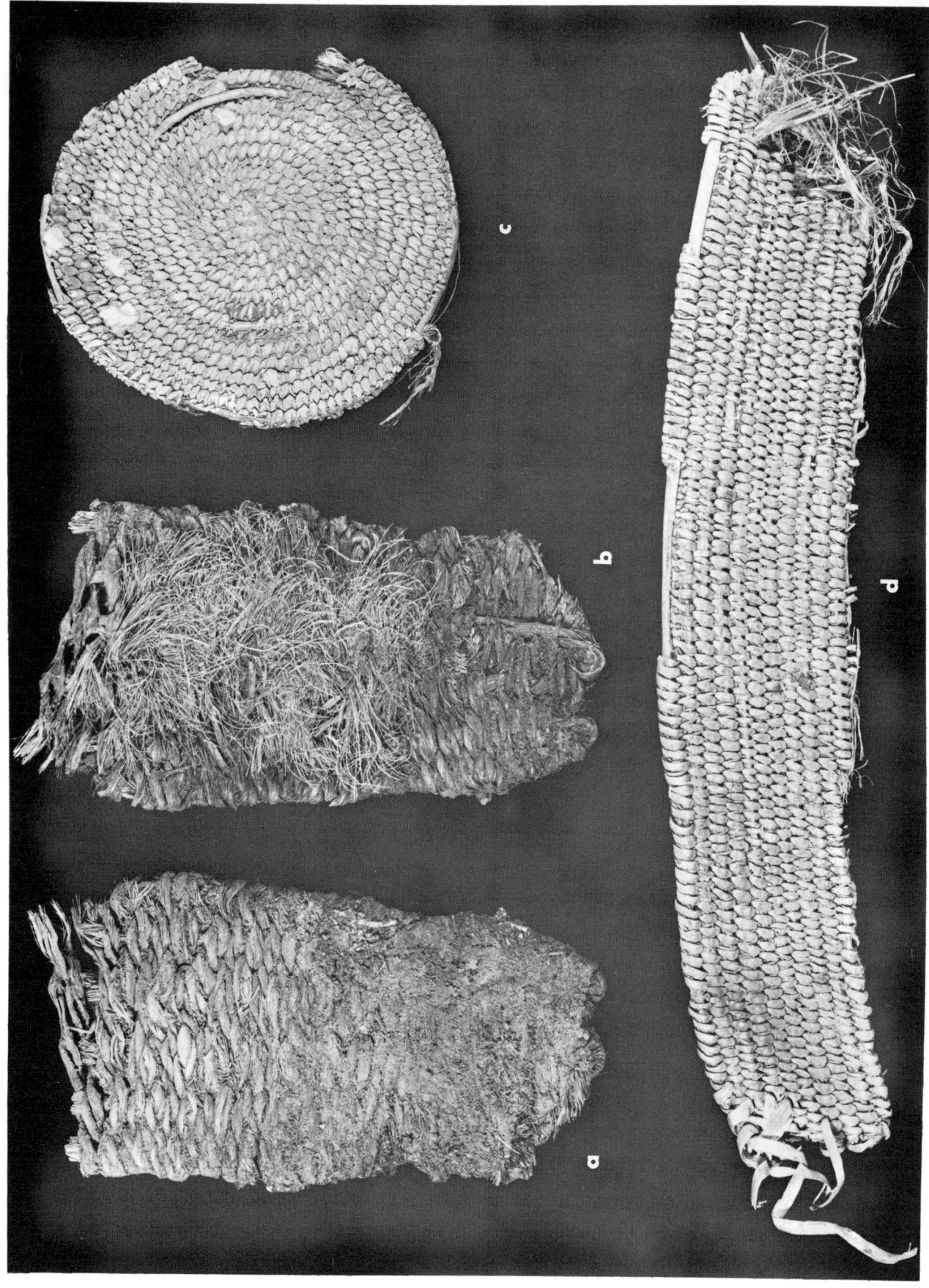

FIG. 29. Fiber artifacts: a, bottom view of cross-weave sandal; b, top view of cross-weave sandal; c and d, examples of one-rod and bundle coiled baskets (note mending stitches on left end of sample d). c is 11 cm. in diameter.

TABLE 14

DISTRIBUTION OF TEXTILES FROM WALTERS CAVE

		I		II		III				IV		V		Sur	TOTAL
Stratum:	NP	A	B	A	B	A	B	C	D	A	B	A	B	Sur	
Cordage															
One Ply									1	2					3
Two Ply Z-twist	2		1						5	1					9
Two Ply S-twist									2	1	2	1	1		6
Netting Fragments											1		1		2
Cordage Material															
Apocynum-Asclepias			1						2	1	2	1	1		8
Grass	1								4						5
Yucca	1								1						2
Juniperus										1					1
Artemisia												1			1
Basketry (Coiled)															
Subclass 2											1				1
Subclass 3												1			1
Subclass 6													1		1
Sandals															
Plain Weave									2						2
Worked Fibers (Misc.)															
Coiled Fibers									1						1
Elongated Splint Rings			1												1
Bent Twigs			1												1
Yucca Strips	1								1			2			4

results; if the spindle is rolled toward the body, the yarn will be Z-twisted. This would mean that in order to make a two-ply Z-twist cord, the first yarn would be rolled away from the body to get the S-twist, then the ply twist would be achieved by rolling the yarns toward the body. Spinning along the leg can also be achieved without the aid of a spindle.

Another method using a spindle is to drop the spindle and let it spin freely just above the ground. Here, as before, the twist direction is determined by the direction the spinner twists the spindle as it is dropped. It is, of course, impossible to say what method was employed by the inhabitants of Cowboy Cave.

The provenience of the cordage is shown in Tables 13 and 14.

NUMBER OF PLIES. Of the 483 total cordage specimens, 437 are two-ply, obviously the most popular manufacturing technique. Of the remaining cordage, 39 are single ply, which may represent cordage that has come unplied (see Table 13 for provenience). One specimen, a two-ply Hawser (SS--S-twist over an S-twist), came from Stratum IVa. A Hawser two-ply is different from a regular two-ply in that the single yarns are twisted in the same direction as the ply twist.

Six multitwist (i.e., more than two-ply) pieces of cordage were recovered from Cowboy Cave. Five of these are semi-Hawser twists, having the first yarns S-twisted, the two-ply Z-twisted, then Z-twisted again with another two-ply Z-twist, creating a four-ply cord. The sixth multitwist specimen is a four-ply cable, the consecutive twists alternating Z-S-Z. Material used for multitwist cordage include *Apocynum-Asclepias* (1); grass (4, 3 of which are *Sporobolus*); yucca and grass (1).

DIAMETER OF CORDAGE. Although cordage dia. ranges from 1 mm. to more than 10 mm., the bulk of the specimens have a dia. between 2 and 3 mm. Table 15 shows that there is some relationship between the diameter of the cordage and the type of material employed. The finer pieces were made from the inside fibers of *Apocynum-Asclepias*, whereas

the coarser cordage was made from grasses, bark, and yucca.

Braided Cordage

Of the five **braids** recovered at the cave, three are rather coarse three-strand **braids** made of yucca. The other two are 10-strand braids (or flat sennit). Of these, one is made of *Apocynum-Asclepias* (no provenience); the other, from Stratum Vb, is made from what appears to be human hair. Both have a dia. of approximately 5 mm. Their specific use is unknown.

Doubled-back Cordage

Fifteen pieces of cordage were doubled back on themselves, then twisted to create a two-ply cord. It is possible that this was the normal procedure in making two-ply cordage; in fact, a spinning experiment in the laboratory showed that a single spun yarn will automatically twist back on itself when the tension of the spindle is released. Due to the fragmentary condition of the specimens and their obvious repeated use, it is indeed possible that the ends showing this manufacturing detail have been broken off. The proveniences show that this technique is continuous throughout all the cultural units.

Knots

Six varieties of knots were recognized in the cordage from Cowboy Cave. Although we do not know whether knot types reflect cultural preferences, the various types have nevertheless been tabulated and their distribution by stratum is represented in Table 13. The occurrence of each type is apparently related to its specific intended purpose. The sheetbend was the most desirable knot for netting, though lark's-head knots were also popular for this purpose. When joining two cords together, which was a very common practice, a square knot was best. Granny knots were also used for joining purposes. Many overhand knots in the ends of cords probably prevented the fibers from unraveling. Table 13 shows that the sheetbend is the most common knot, the overhand second, and the square knot third. It has been noted that the square knot

TABLE 15

CORDAGE MATERIALS IN RELATION TO CORDAGE DIAMETER*

Diameter mm.	Apocynum-Asclepias	Grass	Yucca	Artemisia	Linum	Juniperus	Cowania	Gossypium	Hair	TOTAL
<1	67	13	3	2	1	2				88
1-2	134	26	5	3	12	1			1	182
3-4	53	21	10	4		1	1			90
4-5	23	12	6	2				1		44
5-6	5	12	4	5					1	27
6-7	4	2		1						7
7-8	1	5	3	1						10
8-9	1	6	1							8
9-10		1								1
>10	1	12	4							17

*Cordage from Walters Cave included.

seems to predominate in the Southwest (Basketmaker through Pueblo), whereas the sheetbend and overhand knots are more common in the Basin (Lambert 1961: 57).

Cordage Artifacts

Netting Fragments

Cordage artifacts identified as netting fragments total 12 from Cowboy Cave and 2 from Walters Cave. All specimens are made from fine (average dia.: 1.4 mm.) two-ply Z-twist cordage made from Apocynum-Asclepias. The mesh ranges from 4 to 12 cm. The largest net fragment was tied in a large overhand knot at one end, then twisted, apparently for storage. This specimen may, in fact, be a complete, though small, net, but it is too fragile to untie. Another smaller fragment was tied in three overhand knots, also probably for storage purposes. Table 13 shows the type and number of knots used for netting manufacture. Undoubtedly, the sheetbend knot was the most desirable for this purpose. Netting fragments occur in Units III, IV, and V, but are most numerous in the middle levels of Unit III.

String Skirt or Apron

The fragmentary remains of a string skirt were recovered from Stratum IVd. They consist of two strands of two-ply S-twist cordage making a belt or draw string, over which strings are draped (also two-ply S-twist); 64 of the strings are still intact. They are actually doubled over the waist cord and held in place by two cords running below the belt cord. These holding cords encircle each consecutive group of three to four longitudinal cords. There are two overhand knots tied in one end of the waist cord. The ends of the apron cords are burned, so that the longest piece is only 10 cm. The waist cord measures 13 cm. The entire skirt is constructed of fine quality 2 mm. *Apocynum-Asclepias* cordage (see **Fig.** 30b). A similar string skirt is reported by Kidder and Guernsey (1919:157, Pl. 66a).

Necklace Fragments

Two small fragments from two different necklaces were found in Cowboy Cave. Both consist of seeds strung on cordage. One, recovered from Stratum Vd, is a single *Scirpus* seed strung on a 4 cm. cord of *Apocynum-Asclepias*. The other fragment, of unknown provenience, consists of seven *Lithospermum ruderale* seeds tightly strung on a 3 cm. cord.

Twined-grass Mats

Two cordage artifacts appear to be portions of twined-grass mats. The warp elements are consistently two-ply Z-twist grass fiber cordage. These elements are bound together with rows of S-twist twining. The twining elements are yucca strips. One specimen, from Stratum IIIg, has a single row of twining; the other specimen, from Stratum IIIf, has three rows of twining.

Miscellaneous Cordage

Three unusual pieces of cordage are of unknown function. The first of these is a piece of grass and yucca two-ply S-twist cordage doubled back and Z-twisted on itself, but in four radiating segments (see Fig. 31d), then two opposite radiating segments were Z-twisted together, giving it the appearance of a four-ply cord. The provenience of this fragment is Stratum IVc. A very similar cord was found in Zion National Park (Schroeder 1955:151, Fig. 23, p. 153).

The second unusual cordage artifact is a mass of five to seven two-ply S-twist cords knotted together at the ends with overhand knots. In addition to the two overhand knots, there are 12 lark's-head, one sheetbend, and two square knots (see Fig. 31a, b). It is possible that this specimen is a badly deteriorated net that was tied with the overhand knots for storage, the same as some of the other nets. It was recovered from Stratum IIIf.

The third piece of cordage placed in the unusual category consists of two pieces of two-ply S-twist treated as one cord, doubled back and loosely Z-twisted together. This gives the appearance of a four-ply cord, but the twist is too loose to have functioned as such and, in fact, seems only temporary (Fig. 31c). The material is *Apocynum-Asclepias*; the provenience is Stratum IVc.

Cordage Materials

According to the identification of plant fibers by Beverly Albee (Utah Museum of Natural History), five major and two minor plants were exploited for the manufacture of cordage materials. The total number of cordage fragments made from each is as follows: *Apocynum-Asclepias*, 283; Grass, 108; Yucca, 44; *Artemisia*, 17; *Linum*, 13; *Juniperus*, 4; *Cowania*, 1 (Table 13). (*Apocynum* [Indian hemp] and *Asclepias* [Milkweed] are so similar in structure that they cannot readily be distinguished, and are therefore classified as one fiber.) Of the grass specimens, 12 could be identified as *Sporobolus* (a genus no longer in the area, perhaps because of overgrazing), two as *Stipa*, and two as *Hilaria jamesii* (galleta grass). Seven of the yucca pieces were identified as the species *baccata*, two possibly *harrimaniae*, and two from the flowering stem of the plant. As shown in Table 13 all of the *Linum* samples were found on the surface, indicating that the exploitation of this plant came quite late in the occupation of Cowboy Cave. All of the *Juniperus* pieces are single ply, the significance of which is unknown. One small fragment of finely spun *Gossypium* (cotton) was found on the surface, but due to the quality and nature of the fiber, it is believed to be historic.

One cordage material category is not of botanical derivation, but human hair. Only two cordage fragments were found manufactured from this readily available resource, one of which was a 10-strand braid (or flat sennit). In addition, there are 21 pieces of cordage made from various combinations of fibers. These

FIG. 30. Fiber artifacts: a, hairbrush (approx. 27 cm. in length); b, string apron; c, bark-wrapped ring.

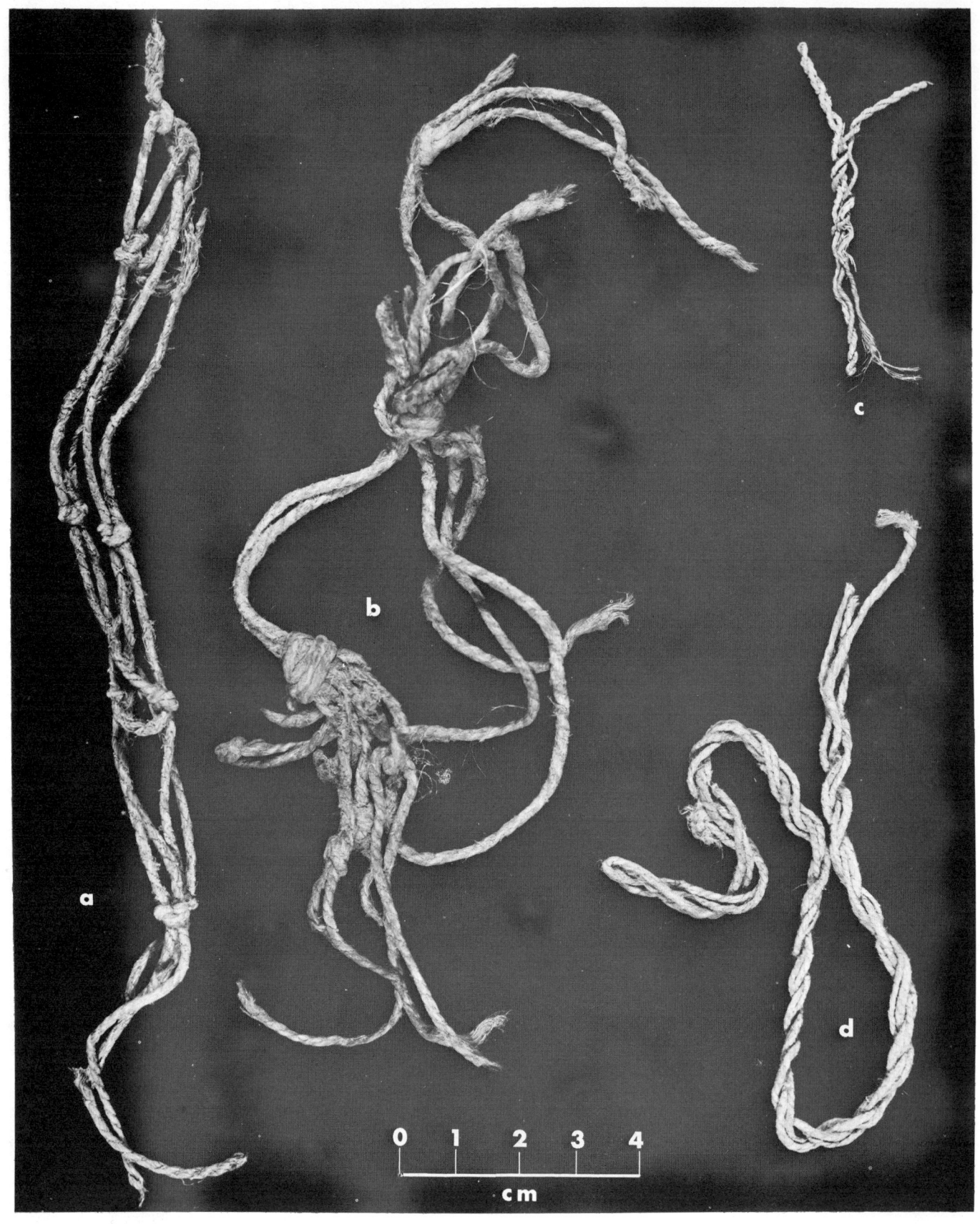

FIG. 31.   Unusual pieces of cordage of unknown function.   a and b are from Stratum
IIIf, c and d from IVc.

[70] AP 104/UU

include: grass and yucca; grass and *Apocynum-Asclepias*; grass and *Juniperus*; *Apocynum-Asclepias* and yucca; *Apocynum-Asclepias*, yucca, and grass; and two specimens of *Apocynum-Asclepias* tied to a cord of animal fur. Their distribution within the site is shown in Table 13.

As can be seen from Table 13, most of the cordage was made from *Apocynum-Asclepias*. This may be due to several factors, the first being the natural breakdown of the internal pulp of these plants into long fibers. It is also possible that these plants made the strongest fibers, since they were always chosen for net making. Availability of the plants may have been another factor involved. Second in popularity was grass, which was apparently quite abundant during the occupation of Cowboy Cave. This fiber, however, was probably not as strong as *Apocynum-Asclepias*. Most of the grass specimens recovered are extremely fragile and shredded. Choice of fibers, however, was probably very dependent upon the intended use of the cord.

WORKED PLANT FIBERS

Artifacts in this category include grass and bark pads, fiber bundles, bark torches, fiber coils, wrapped rings, elongated splint coils, buttons of wrapped splints, bent twigs and splints, yucca strips and knots, problematical objects, and yucca strips and bundles.

Grass and Sagebrush Bark Pads
Fifteen shredded grass pads and two sagebrush bark pads were recovered from Cowboy Cave (see Table 12). The size and shape of several of these pads suggest that they might be innersoles of sandals. In fact, nine of the sandals in the collection still retain virtually identical innersoles. Five of these pads are caked with mud.

The remaining pads (7), are smaller and rounder. Aikens (1970:119) has suggested that they might represent menstrual pads. However, a variety of

tests for human blood on the pads were entirely negative, so there is no conclusive evidence of their purpose. Two of these pads are slightly charred. As a whole the fibers are compressed, sometimes stuck together.

Shredded Sagebrush Bark Bundle
A large hank of very finely shredded sagebrush bark is loosely twisted and self-wrapped into a bundle. This specimen is nearly identical to one recovered from Hogup Cave (Aikens 1970:119). It comes from Stratum IVc.

Fiber Bundles
This category is a heterogeneous group of bundled fibers which differ in size, material, and general appearance; there are 16 in all. Three are small grass bundles, and three are small fiber-wrapped bundles of twigs. Eight are variously sized, wrapped or twisted bark bundles, one held together by strips of yucca tied in square knots. One large bundle is a combination of grass and bark, and is loosely wrapped. One bundle of juniper bark 70 cm. long and wrapped most of its length with another piece of bark may be a rope. Their proveniences appear in Table 12.

Bark Torches
Two long (18 cm. and 36 cm.), compact bundles of juniper bark tightly wrapped with thin stick splints may have been used as torches. Both specimens are charred at one end; one is from Stratum IIIf, the other has no provenience.

Coils
Most of the coiled fibers of this category probably were used to bind or wrap other fibers or objects together. This is a very diverse group of artifacts in size, shape, and material. However, only a few of the coils have knots and, therefore, were probably secured as binders by tucking under the loose ends. Several of these pieces may have been the stitching elements from coiled baskets. Similar coils are reported from Sand Dune Cave (Lindsay et al. 1968:85). Their proveniences are shown in Tables 12 and 14.

Wrapped Rings

Six fiber-wrapped rings were recovered from Cowboy Cave. The two smallest wrapped rings measured ca. 3 cm. in dia. A split-rod ring, measuring about 4.5 cm., was fiber-wrapped only where the ends of the rod met. Two rings are made of bark and wrapped with the same material. The sixth specimen is a stick ring wrapped with bark; it measures 11 cm. in dia. (see Fig. 30c). Artifacts of this size and appearance reported from other sites are thought to have been rings for a hoop and dart game. Table 12 shows their proveniences.

Elongated Splint Rings

This group of artifacts comprises 19 elongated and flattened coils of flat split twigs or splints. These splints are virtually identical to the splints used for the sewing elements in coiled basketry. However, these artifacts are small, averaging 5 cm. in length and 1.5 cm. in width, and do not appear to be connected with basketry construction. All of these rings are made of single splints and are tied around the middle by a loose end or another fiber. Figure 32 is a photograph of an elongated splint ring. Artifacts of similar construction are known from Dust Devil Cave (Lindsay et al. 1968:115), and reported by Kidder and Guernsey (1919:144). However, these specimens are double; that is, two elongated rings interlace each other at right angles, and have a handle on one end. According to Lindsay et al. (1968:115), these objects resemble those used by the modern Navajo as toys hung on cradle boards. Table 12 shows the provenience of the Cowboy Cave rings, Table 14 the single example from Walters Cave.

Buttons

This group of artifacts consists of nine tightly coiled circular "buttons" of splints. The coils are held together by splint elements passing through the center of the coil, wrapping around the outside of the coil, then coming back through the center. The stitching is continued in this fashion around the circumference of the coil, creating a small, button-like object (see Fig. 33c, d). Although the technique of coiling and stitching is similar to coiled basketry, each of these artifacts seems complete as it is. The average dia. of the artifacts is 2.5 cm. No speculation as to the use of these objects can be made and no comparable objects have been reported. See Table 12 for their distribution.

Bent Twigs and Splints

Five variously bent or knotted twigs and one accordian bent splint may represent a kind of doodling. Also included in this category is an unidentifiable object consisting of two split rods alternately bent and wrapped around each other, creating a series of false knots. This object is 6.5 cm. long and 1 cm. wide. It was found in Stratum IVc. The proveniences for the bent twigs are shown in Tables 12 and 14. Other bent twigs are reported in *Wood and Reed Artifacts*.

Problematical Objects

Among the Cowboy Cave collection are 11 examples of an unidentifiable object which consists of a split rod or splint of *Rhus trilobata* (squawbush) doubled back on itself in order to be self-wrapped. At each end wrapping, the splint is split again for a few centimeters, creating a fringe along one edge (see Fig. 33a, b). There is no evidence to show how these objects may have been incorporated into a more complete object. Nor is there any evidence to suggest their attachment or use with another object. The average length is 10 cm.; they were recovered from Units IV and V (see Table 12). Objects nearly identical to these were reported by Morss in the Fremont River Area (1931:74). Hurst reports these same objects from three sites in southwest Colorado. Two were recovered from Dolores Cave (1947:13, Pl. II, Figs. 27 and 29), which is essentially a Basketmaker site; one from Tabequache Cave II (1944:9, Pl. II, Fig. 2) also defined as of the Basketmaker culture; and a fourth similar artifact came from Cottonwood Cave (Hurst 1948:12), a "peripheral" Basketmaker site.

Yucca Strips and Knots

This large group of artifacts consists of yucca leaf strips which were used to bind construction materials together, to tie up bundles, or as sandal ties, among other uses. Although the yucca fibers are **strong**, once the strips dry they become brittle and are easily broken. This probably accounts for the large quantity of these artifacts. Out of a total of 251 yucca strips, 102 are plain strips, while the remaining 149 are knotted, some more than once. The latter include: 127 square knots, 28 overhand knots, 7 granny knots, 4 lark's-head knots and 11 unclassified knots. Most of the knots joined two strips, the square knot being overwhelmingly preferred. The provenience of yucca strips from Cowboy Cave appears in Table 12, and from Walters Cave in Table 14. Over 50 percent of these artifacts are from Unit III.

Yucca Leaf Bundles

One large bundle of approximately 50 unprepared yucca leaves, loosely tied by a yucca strip binder, was recovered from Stratum Vb. A smaller bundle of thin, folded, prepared yucca leaf strips was found in the test trench during the 1973 work. These bundles were undoubtedly raw materials gathered for future use as binders or in sandal construction.

Hairbrushes

Two bundles of rigid grass stems of uniform length are believed to be hairbrushes. The utilized ends of both brushes contain short hairs and are stained. These artifacts are constricted in the middle and are uniformly twisted in a manner that causes the ends to flare slightly. Both brushes are tied around the middle with pieces of fine two-ply S-twist cordage, one with a single square knot, the other with one lark's-head and seven sheetbend knots. A similar brush is described by Lipe (1959:224, Fig. 53) (see Fig. 30a). They come from Strata Vb and Vc.

FIG. 32. An elongated splint ring, 4.4 cm. long.

FIG. 33. Fiber artifacts: a,b, problematical basketry objects; c,d, problematical "button-shaped" objects. c is 2.5 cm. in diameter.

WOOD AND REED ARTIFACTS

Joel C. Janetski

INTRODUCTION

Cowboy Cave, which was totally dry, contained some 173 well-preserved wooden artifacts. Their identification as artifacts was based on any occurrence of smoothing, cutting, shaping, or binding of the specimens. These criteria excluded from the discussion the many charred and scorched sticks found in the various strata, although these were probably associated with human activity. Once identified as cultural, the artifacts were segregated on the basis of function, although, where use was not apparent, descriptive categories were established. Some limited identification of the woody plants was made. The artifacts are described below and compared with similar recorded finds in the Great Basin.

Atlatl

A smoothed and polished (although slightly warped) atlatl (Fig. 34m) was found in Unit IV near the east side of the cave some 7.5 m. from the entrance. The atlatl, made of straight-grained mountain mahogany (*Cercocarpus*), is broken and charred at both ends, although significant features fortunately remain. About 2 cm. from the proximal end the atlatl has been roughly and abruptly notched. Lateral scoring extends from the notches toward both the distal and proximal ends for ca. 2 cm.

The beginning of a central groove 2.8 cm. from the distal end suggests that this was a female atlatl, according to the distinction made by Hester et al. (1974: 7). Its overall measurements are 42.2 cm. long by 2.1 cm. wide by 0.9 cm. maximum thickness. In cross section it is a flattened oval.

The notches on this atlatl are unlike the more gentle indentations cut out to accommodate a finger grip, as depicted in Lindsay et al. (1968:65, Fig. 41), Hester et al. (1974, Pl. 2), Kidder and Guernsey (1919:180), and Guernsey and Kidder (1921: 80-83, Pl. 33). They are more similar to the notches on the proximal segment described by Dalley (1970:154) from Hogup Cave, although, as Dalley points out, that segment also shows evidence of the beginning of a constriction immediately above the notches, which no doubt was for the finger grip. Dalley suggests the notches are in some way related to securing the separate finger grip apparatus to the atlatl.

Because of the absence of the rounded finger indentations, the lack of frontal flattening, and the abrupt notching, some consideration was given to classifying this artifact as a bow fragment. However, there was no wear in the notches on the front or back of the artifact, which would be expected from the use of the bow string. Also, the notches were cut at right angles to the shaft, rather than at a diagonal, as they appear on a previously

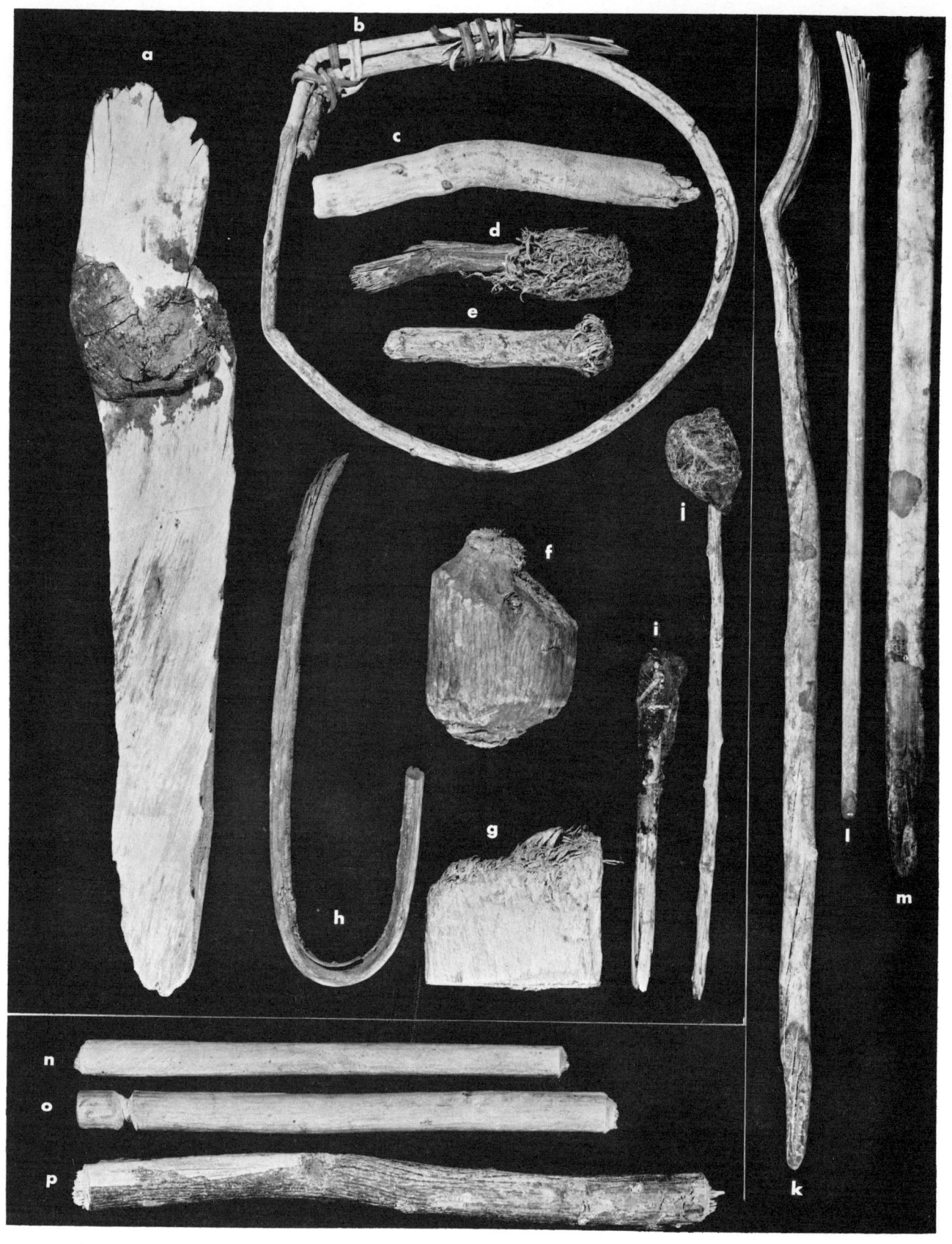

FIG. 34. Worked wood: a,i,j, sticks with pitch; b, wooden hoop; c-g, woodworking waste; h, bent stick; k,l, digging sticks; m, atlatl; n-p, cut wooden shafts.

collected bow fragment (Gunnerson 1959: 79, Fig. 28a). This specimen also differed from other described bows in cross section. In Cosgrove (1947:61) bows were all either round or D-shaped throughout their length; this is a flattened oval. In addition, there was an incipient groove at the charred distal end. Finally, the specimen was stratigraphically contemporary with dart shafts and underlay the arrow parts, which were confined to Unit V.

Atlatl Dart Shafts

Of the various wooden shafts found at the site, 12 were labeled atlatl dart shafts. This identification was based on size--over 0.9 cm. in dia., after Grosscup (1960) and Dalley (1970); and, in the case of six of the sections, on recognizable distal or proximal ends of the shafts. The other six sections can only be tentatively assigned to the dart shaft category on the basis of the two former criteria. As can be seen from the provenience chart (Table 16), four shafts are from Unit V, five from Unit IV, one from III, and two have no provenience. Wood materials include *Cornus*, *Populus*, and *Cercocarpus*.

MAINSHAFTS. Three broken shafts were identified as the proximal, central, and distal portions of mainshafts of composite atlatl darts. The first of these is a smoothed fragment which tapers slightly to the cupped proximal end. It measures 7 cm. in length and is 0.75 cm. in dia. at the cupped end and 1 cm. in dia. at the broken end. It appears that the shaft is broken immediately above the fletching because it is lighter in color just above the break, as though it had been wrapped while in use.

The central portion has two wrappings of sinew 12 cm. apart; it is broken just beyond each wrapping. The sections of the shaft between and below the sinew are stained dark red. This fragment measures 20.5 cm. overall and 1 cm. in dia. .

The distal section is split most of its length; it is smoothed and shaped and socketed with a hole 0.7 cm. in dia.

and 3 cm. deep. The surface has been rasped and roughened for 2.5 cm. The fragment measures 30.3 cm. long by 1.5 cm. in dia. at the unbroken end.

DART FORESHAFTS. Foreshaft sections include one proximal end, one distal end, and one complete wooden point (Fig. 35b). The proximal fragment is smoothed and polished and its proximal end has been rasp-roughened for 3 cm. to a rounded point; the other end is charred. Total length is 19.5 cm. and the dia. is 1.1 cm. throughout. The distal fragment is tipped with black pitch mixed with grass which was apparently used to secure a projectile point. The shaft is smoothed and shaped and is scarred for 1.2 cm. below the pitch, where it was probably wrapped with sinew for additional strength. The wooden point is scraped and whittled to a rough torpedo shape. The blunt end of this wooden point is stained dark red. It measures 8 cm. long by 0.85 cm. in dia. at its widest point.

BROKEN SHAFTS. There are six additional smoothed and rounded shaft fragments; these are either broken or charred on each end. They range in dia. from 0.9 cm. to 1.3 cm. They are most likely dart shaft sections, although, with one exception, they lack changes in coloration or scarification associated with sinew wrapping. The exception, from Stratum IIIi, is especially well shaped and polished, and does appear to be darkly, although irregularly, stained.

Atlatl dart shafts are common in dry cave sites in Utah, Nevada, Arizona, and New Mexico.

Arrow Shafts

Seven mainshafts of *Phragmites* reed and two wooden foreshafts, all from Unit V, were judged to be parts of composite arrows. All are crushed or broken to some extent, although three can be tentatively identified as proximal sections, two possibly as distal sections, and one as a fletched section of a mainshaft.

The most complete of the reed mainshafts (Fig. 35d) is a proximal section, V-notched and sinew wrapped just above a joint. It measures 9.2 cm. long by 0.6 cm. in dia. The second of the proximal

TABLE 16

DISTRIBUTION OF WOODEN AND REED ARTIFACTS

Unit:	NP	Test Trench				I		II		III											IV				V				Sur	TOTAL
Stratum:		1	2	3	4	a	b	a	b	a	b	c	d	e	f	g	h	i	j	k	a	b	c	d	a	b	c	d	Sur	
Atlatl																								1							1
Dart Shafts	2																		1					1	4		3	1			12
Arrow Shafts	1		1																								4	1		1	8
Snares	1																							1			1	1			4
Fire Hearths																			1											1	2
Awl																											1				1
Gaming Pieces	1																										3		1		5
Drilled Wood	2																										1				3
Wooden Cup																								1							1
Shafts with Cut Ends	2																	1	1					1							5
Shafts with Tenons	1								1																	1		2			5
Digging Sticks	1														1											2	2			1	7
Wooden Pegs	1								1				1																		3
Sticks with Pitch																											4	1	1		6
Woodworking Waste	4							1								1		1			1				3	1	4				16
Wooden Hoop																								1							1
Bent Stick																											1				1
Bound Twigs	2				1										1												3	1		1	9
Split Wood														1									1					1			3
Misc. Worked Wood	6					1									1		2	1	1	1			2	2	1	6				24
Decorated Reed	1																										2			3
Reed Sections	2																										2			4
Reed with Fiber Knots																	1						3	2		1	1			8
Worked Reed	4													1			1						2	1	2	8				19
Split-twig Figurines	5																						5	5	3	3		1	22
TOTAL	36		1		1	1		1	2				1	2	3	1	5	5	1	1	1	1	18	17	10	45	13	2	5	173

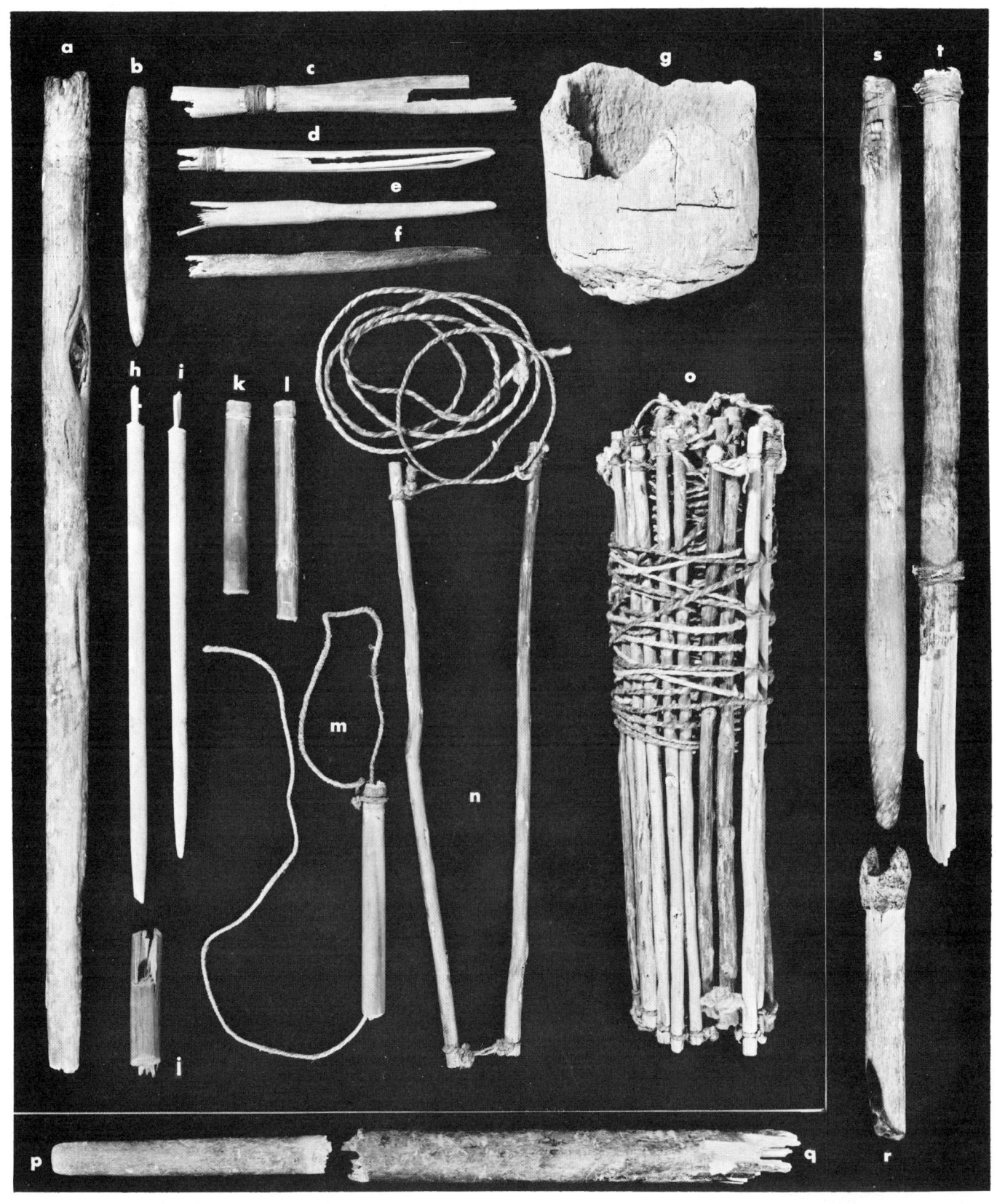

FIG. 35. Worked wood and reed: a,b,p-t, atlatl shaft sections; c-f, arrow parts; g, wooden cup; h-j, shafts with tenons; k-m, noose snares; n,o, scissor snares.

fragments is charred throughout its length, but a portion of the nock and the scarification of the reed for sinew wrapping are still evident. It contains a fitted wooden plug which was inserted into the nocked end of the mainshaft, probably to prevent crushing of the reed when the arrow was being drawn, as described by Martin et al. (1952:340) and Cosgrove (1947:62, Fig. 75). In this fragment, the nocking occurs on the wooden plug rather than on the reed. The remaining proximal section, which is unburned, also contains a plug. Though the wood used for plugs was not firmly identified, it seems quite soft and is believed to be *Populus*. These latter two mainshaft sections measure 4 cm. by 0.8 cm. and 5 cm. by 0.75 cm. in length and dia., respectively.

A reed fragment from the fletched section of a mainshaft is crushed and splintered, but retains sinew wrapping with feather remnants still bound at two points under the sinew. This fragment measures 14 cm. long by 17 cm. in dia. at the wrapping, which occurs at a joint. Another fragment, this one of wood, and measuring 3.7 cm. by 0.5 cm., is also sinew wrapped and contains feather fragments; most likely it is part of a wooden mainshaft.

Two fragments are tentatively identified as distal ends of arrow mainshafts, although they could well be parts of other compound shafts, e.g., fire drills. Both are reamed out internally for ca. 2 cm., possibly to receive an extension of some sort. One is quite short (2 cm.), is broken just above a joint, and has been scratched for 1 cm. above the socketed end. The other measures 13.8 cm. long, is smaller in diameter than any of the other sections (0.5 cm.), has been well scarified for 1 cm. immediately above the socketed end, and may have been cut off at the other end at a joint.

The identification of two wooden objects as arrow foreshafts is also tentative, as both are broken and one is charred. The charred shaft has been smoothed and tapered by a rasping process to the pointed end where the

charring occurs. It measures 9 cm. long by 0.6 cm. in dia. at the broken end. The other shaft is broken at what is probably the distal end, and the proximal end has been rather abruptly reduced from 0.7 cm. in dia. to 0.5 cm. at which diameter it continues for ca. 5 cm., when it comes to a point. The finish work on this end is rather rough, though not the rasp-roughened effect seen on the atlatl foreshaft. Total length of the shaft is 9 cm.

Wood and reed parts of composite arrows are commonly found in Southwest sites where perishable artifacts are preserved. An example of an arrow mainshaft with the foreshaft in place is shown in Haury (1975, Pl. 36). Proximal ends of such shafts are also seen in Martin et al. (1952:390, Fig. 14c-g), Cosgrove (1947, Fig. 76), and Lindsay et al. (1968:67, Fig. 42).

Snares

SCISSOR SNARES. A bundle of 23 well-preserved scissor snares was recovered from Stratum Vc of Cowboy Cave (see Fig. 35n, o). Each snare consists of two fibrous *Chenopodium* sticks 18 cm. long which are notched about 0.5 cm. from each end. At one end the sticks are linked together with cordage tied in the notches with overhand knots, so as to leave about 2 cm. between the lower ends of the sticks. At the other end, one of the sticks has a short cord loop securely bound with sinew wrapping; the other stick has a cord some 90 to 100 cm. long firmly attached with another overhand knot. The long cord passes through the sinew-bound loop so that the two sticks are squeezed together when it is pulled. The cord is Z twist which has been tied off with a simple overhand knot at each end to prevent unraveling.

Scissor snares were labeled and their use described in some detail by Spier (1958:3-4). Other documentation of such snares is found in Schellbach (1922), Gunnerson (1959:95), Kidder and Guernsey (1919:92, Pl. 41), Cosgrove (1947:136-7, Fig. 128), and Fowler (1963:66). Variations on the type described above are reported with the Schellbach snares and on one of the Garfield County snares reported by Fowler. On the former the upper ends of the sticks have been tied together with a 12 to 15 cm. length of cordage. A longer cord is tied in the middle of this joining cord, leaving a short dangling piece with a pointed twig tied to the end of it. Spier (1958:4) speculates that this pointed twig served as a baited trigger. The Fowler variation is identical to the Cowboy Cave artifact except that, at Cowboy, one of the snares has a 3.4 cm. by 0.3 cm. pointed twig tied to the end of the long pull cord which suggests some variation in use.

According to the model documented in ethnographic field notes by Spier (1958, Fig. 1) the use of scissor snares requires seven additional worked wooden objects, plus the "bent twig" energy source per snare. In order to set this bundle of 23 snares, then, over 160 additional functionally specific wooden items would have to be fashioned. Since none of these were found at Cowboy Cave, there may be a simpler method of setting the snares than the one shown by Spier.

NOOSE SNARES. Another type of snare found at Cowboy Cave is a noose snare made of a section of *Phragmites* reed and cord. A complete snare of this type is seen in Figure 35m. The cord is attached just below a reed joint which keeps it from slipping off the end of the tube. After being tied off, the cord passes down through the reed tube. This tube measures 7 cm. by 0.7 cm. in dia., and the two-ply S-twist cordage is 64 cm. long; it has no provenience.

Two cut and notched hollowed-out reed sections which are speculated to be loop snares without the associated cordage, are also illustrated (Fig. 35k, 1). They measure 6 cm. by 0.7 cm. and 6.5 cm. by 0.7 cm., and come from Strata Vb and IVc, respectively.

Identical noose snares are on display in the museum at Mesa Verde National Park. Although they are identified there as snares, the way in which they were used is not depicted. A similar, though somewhat longer, snare was excavated by Burgh and Scoggin (1948) in Mantles Cave in the Uinta Basin. Etna Cave (Wheeler 1942) also contained similar artifacts, although the Eskimo snare Wheeler exhibits in Figure 26 is much closer morphologically to the Cowboy and Mantle Cave types.

Fire Hearths

Two fire hearth fragments were found in Cowboy Cave (see Fig. 36t, u); one was encountered on the modern surface of the cave, while the other was in Stratum IIIi. Both are made of soft, woody materials which could not be identified. The surface hearth is 2.5 cm. long by 1.7 cm. wide by 0.9 cm. thick and shows evidence of fire-making activities on both sides. The other hearth, which has been split lengthwise through the holes, measures 9.3 cm. long by 1.8 cm. wide by 0.6 cm. thick and shows four drilling scars. The surface hearth has one complete drilled hole 0.7 cm. in dia. which has the characteristic slot in the side to allow the embers to drop onto the tinder material.

Awls

One slender twig (see Fig. 36v) from Stratum Vb is shaped and tapered into an awl. It measures 8.5 cm. long by 0.3 cm. in dia. at the butt end. The point appears to have been sharpened through rubbing or grinding. The material is probably *Chenopodium*.

Gaming Pieces

Five flat wooden items were classified as gaming pieces (see Fig. 36e-i). Two of these are decorated, while the others are either unfinished or finished but blank. The most elaborate of the decorated pieces is a smoothed wooden rectangle 2 cm. long by 1.2 cm. wide by 0.2 cm. thick with seven regularly spaced straight cuts across its width. The piece was cut from the surface of a branch, and the incisions appear on the naturally smooth side. This piece, found in Stratum Vd, has been identified as *Juniperus*. The other decorated piece, from Stratum Vb, is also a rectangle, though not as nicely smoothed as the first. It has been incised four times

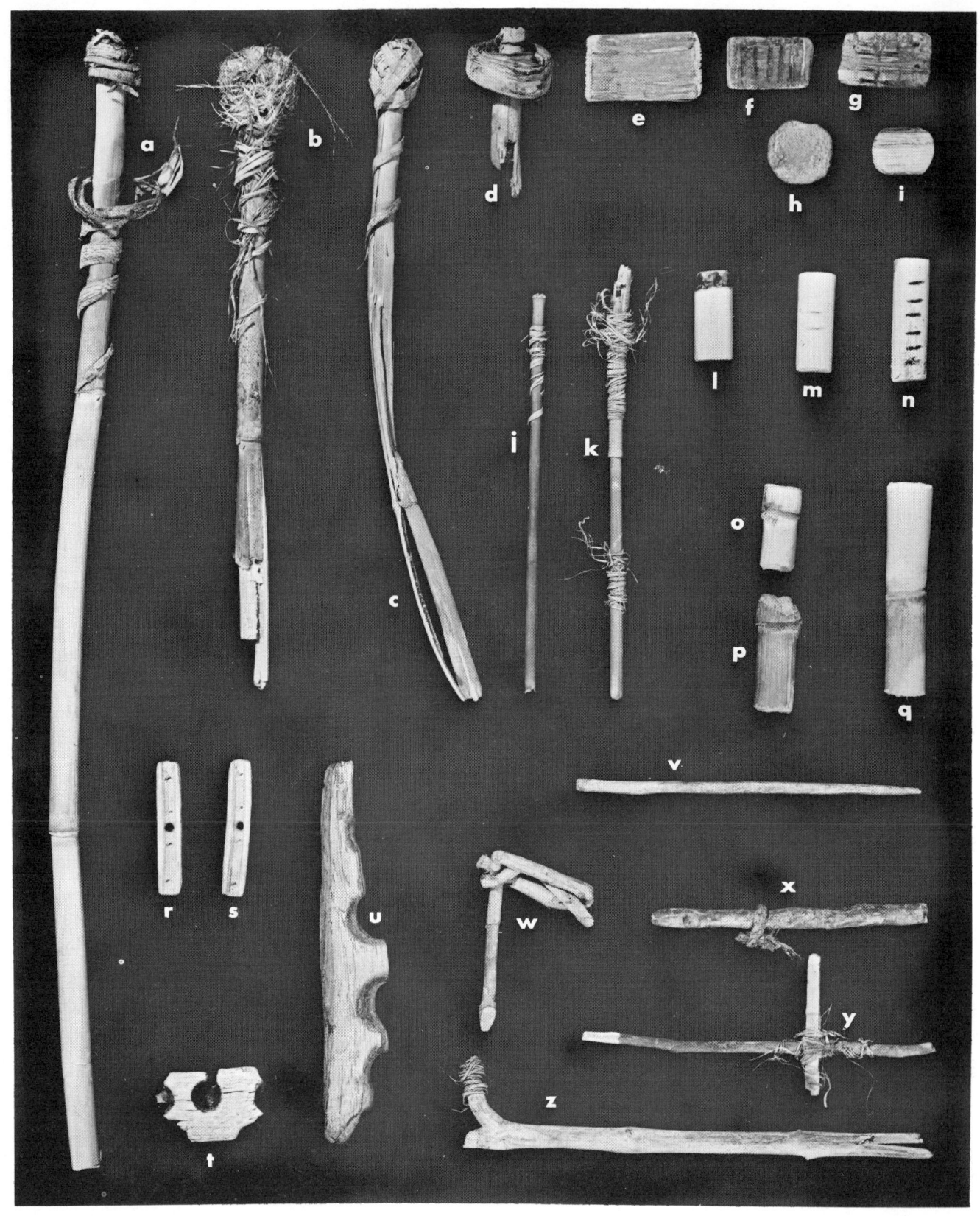

FIG. 36. Worked wood and reed: a–d, problematical objects; e–i, gaming pieces; j,k,w–z, bound twigs; l–n, decorated reed; o–q, reed sections; r,s, drilled objects; t,u, fire hearths; v, awl.

on the rough side. It measures 2 cm. long by 1.2 cm. wide by 0.2 cm. thick. In both specimens the incisions are partially filled with charcoal, probably to emphasize the decorations.

A wooden rectangle identified as *Juniperus* measures 2.7 cm. long by 1.5 cm. wide by 0.25 cm. thick. It also has been cut from the surface of a branch and has one smooth and one rather rough side. It has no provenience.

A smaller, though thicker, piece is also basically rectangular but has had the ends rounded off. It measures 1.4 cm. long, 1.1 cm. wide, 0.35 cm. thick, and comes from Stratum Vb. Neither of the sides of this piece are smoothed, nor is it cut from the surface of a branch.

The last gaming piece, also from Stratum Vb, is a bark scale which has been rounded to a disk about 1.6 cm. in dia. and 0.2 cm. thick.

The only comparable object reported in the observed literature was one item from Dust Devil Cave identified as a gaming piece (Lindsay et al. 1968:114, Fig. 93c). These similar artifacts from Cowboy Cave could be gaming pieces or possibly decorative pieces of some kind, although they are a bit smaller.

Wooden Cup

A section of a soft, porous *Populus* root found in Stratum IVc was roughly hacked off at one end and hollowed out at the other to form a crude cup (see Fig. 35g). The artifact measures 6.2 cm. in total height, and 6.2 cm. in dia. The inside of the cup measures about 5 cm. from the bottom to the highest portion of the broken, irregular rim.

There are two cottonwood cups mentioned by Kidder and Guernsey (1919:121) and four in the Cummings collection at the Utah Museum of Natural History. All have holes drilled in the sides, and are polished and smoothed from manufacture or use. The Cowboy Cave cup, in contrast, is not smoothed, nor does it have a hole in the side. The Kidder and Guernsey cups, although not dated, are associated with structures and are, therefore, probably younger than the Cowboy Cave artifact.

Wooden Shafts with Cut Ends

Five objects were categorized descriptively as shafts with cut ends. One of these is a smoothed shaft which has been deeply notched ca. 1 cm. from one end (Fig. 34o). The notched end has been cleanly cut and worn, while the other is snapped off at a point weakened by cutting around the twig. This shaft measures 12.8 cm. long, averages 0.8 cm. in dia., and comes from Stratum Vb.

Another shaft severed at both ends by cutting around and snapping has not been smoothed in any way; in fact, the bark is still intact throughout its length (Fig. 34p). It measures 15 cm. long, 1.2 cm. in dia., and has no provenience.

A third shaft from Stratum IIIi is smoothed and tapered to a charred end. The blunt end was severed by notching and snapping. It measures 21 cm. long and 1.2 cm. in dia. at the cut end. Another similar shaft which has no provenience is smoothed through use and charred at one end. The other end is apparently cut, but may have been merely broken and the ragged fibers worn down and roughly rounded. It measures 15.7 cm. long and 1.2 cm. in dia. at the cut end.

The last shaft (Fig. 34n), cut by notching and snapping at both ends, has been abraded and smoothed throughout its length. It measures 11.5 cm. long and 0.85 cm. in dia., and has no provenience.

Shafts similar to Figure 34n are seen in Cosgrove (1947:153, Fig. 149) where they are called gaming counters. Also, Dalley reports various shafts with cut ends which he classifies as wooden cylinders (1970:171). Dalley's descriptive term is understandable, as a functional interpretation of these items is difficult in the absence of ethnographic parallels.

Shafts with Tenons

Of the five objects in this group, three are almost identical, varying slightly in length. The three have carefully smoothed shafts which taper to slightly blunted points and the other ends cut off by a method described and depicted by Cosgrove (1947, Fig. 71). The process consists of cutting opposing notches on the stick about a third of the way through, leaving the center intact. Then, after scoring the stick at the desired breaking point and prying on the notch to weaken the fibers, the tab is broken out by simply bending the tenoned portion of the stick.

Two of these similar tapered shafts (Fig. 35h, i) have the tabs still intact, while the third has had the tab broken off. The two shafts with tenons were found together in Stratum Vc and measure 0.55 cm. in dia., and 14.1 cm. and 15.8 cm. in length, respectively. The longer of the two was identified as *Cercocarpus*. The shaft without the tab measures 0.6 cm. in dia., 10.2 cm. long and comes from Stratum Vb.

The remaining two shafts with tenons are considerably shorter and larger in diameter than those just described. One, from Stratum IIb, measures 4.4 cm. long and 0.9 cm. in dia. It is smoothed and is cut on one end by the opposed-notch method, leaving a tab stub. The other end is slotted by the same process. The last shaft, which has a crude point on one end and a wedge-shaped tenon on the other, measures 5.7 cm. long and 0.85 cm. in dia. It has no provenience.

Similar items with tenons have been found in various sites in Utah (Dalley 1978; Dalley 1970:167), as well as in Arizona (Cosgrove 1947), and they are generally considered to be waste produced in notching arrow or dart foreshafts to receive projectile points. It seems unlikely that the three carefully smoothed and tapered shafts would be waste material, though it is possible that they were wooden arrow foreshafts which had been cut off to allow for conversion from wooden to flint points. This might explain the slightly blunted tips on these shafts. A comparable artifact, called an arrow foreshaft, is described by Dalley (1970:167).

Digging Sticks

Seven objects were tentatively identified as digging sticks. One, from Stratum IVa, is a slender, curving stick 57 cm. long and 1 cm. in dia. made of a fibrous chenopod material. Both ends are charred; one is somewhat flattened and split for ca. 12 cm. Another, which was found on the surface, is quite straight except for a crooked end (Fig. 34k). It is also charred on both ends and shows some signs of abrading. It measures 57.3 cm. long and 1.5 cm. in dia. Check cracks on the surface of this specimen were packed with sand mixed with a rather large number of chenopod/amaranth seeds which suggests that this digging stick was also used to beat seed-bearing plants.

A shorter stick with some of the bark still intact is charred on one end while the other has been split back on each side to form a two-sided blade much like that described by Dalley (1970:174). The blade end is worn and some striations are visible. The stick measures 28.3 cm. in length, 1.2 cm. in dia. and comes from Stratum Va.

The most carefully made of these digging sticks is smoothed and polished to a rasped, partially flattened, fire-hardened point (see Fig. 34*l*). It is cracked about 7 cm. up from the tip and the far end is broken and shattered. It comes from Stratum Va and measures 39 cm. long and 1 cm. in dia. Another digging stick, from Stratum IIIf, is a flat, split piece of twisted wood charred at both ends, although one end comes to a flat point and appears to be worn. It measures 44.5 cm. long, 3.8 cm. wide, and ca. 1.2 cm. thick.

The last two artifacts in this category are probable tips of digging sticks. One has a well-worn flattened blade for a point and measures 14.3 cm. long and 1.1 cm. wide. The other has a tapered, abraded point and measures 11 cm. long and 1 cm. in dia. Their proveniences are Strata IVc and Vb, respectively.

Wooden Pegs

Three short, pointed sticks were categorized as pegs. One is quite charred but appears to have been smoothed to a point at one end; the other end is cut off by the opposed notching method. It measures 6.5 cm. long and 1 cm. in dia. at the cut end and was found in Stratum Ib. Another short stick is roughly rasped to a blunt point on one end and cut off quite cleanly by a sawing technique on the other. It measures 8.3 cm. in length and 1.6 cm. in dia. at the cut end and has no provenience. The third peg is longer, 15.8 cm., 1.1 cm. in dia., and is also roughly rasped to a blunt point which is somewhat charred. Its provenience is Stratum IIId.

Sticks with Pitch

There were six sticks found with pitch gobbets on the ends or along the length. Four of these are unworked sticks, one is carefully worked, and one is a *Phragmites* reed. One of the unworked specimens (Fig. 34j) is simply a thin twig 0.4 cm. in dia. and 16 cm. long with a round ball of pitch 2 cm. in dia. on the end. Another of these (Fig. 34a) is a flat piece of wood 25 cm. long, ca. 3.5 cm. wide, and ca. 1 to 1.5 cm. thick with a smear of pitch on the smooth outside surface. The other unworked pieces and the reed measure 7 cm., 9 cm., and 10 cm. long, respectively, and are all under 0.7 cm. in dia. Four of these artifacts are from Stratum Vb and one is from Vd.

The remaining stick with pitch (the worked one) has been flattened and shaped to rounded points on both ends and smeared with pitch after it was integrated into some larger artifact; both flat sides show a pattern of rounded impressions in the pitch, probably from small sticks. This piece measures 7.3 cm. long, 0.6 cm. wide, 0.2 cm. thick and is from Stratum Vc.

Lindsay (1968:71, Fig. 44g) reports what he calls a pitch brush from Sand Dune Cave which is not unlike the pitchy sticks from Cowboy. These may very well have been tools used in caulking baskets.

Woodworking Waste

Some 16 sticks were identified as waste from woodworking activities. Of these, nine are shafts, or at least round sticks of varying diameters and lengths, but all are less than 1.8 cm. in dia. and 10 cm. in length. All but one are smoothed for a distance at one end and then cut off by notching, either around or on opposite sides. The uncut ends are rough, unsmoothed and broken.

Three of the 16 are flat sticks which are cut off and smoothed for a distance above the cut. One is cut, though not smoothed, on the other end; two are rough and frayed. These pieces measure 5 cm. wide, 4.5 cm. long, and 1 cm. thick; 3.3 cm. wide, 7.3 cm. long, and 1.1 cm. thick; 1.7 cm. wide, 19.7 cm. long, and ca. 0.1 cm. thick.

Another waste piece (Fig. 34f) is a stub of *Quercus* (oak) 3.5 cm. in dia. and 6 cm. long, from Stratum IVd. Both ends have been cut off through a cutting and sawing process.

The remaining three waste pieces are sticks burred at one end (see Fig. 34c, d). They measure 6 cm. long, and 1 cm. in dia.; 7.5 cm. long, 0.7 cm. in dia.; and 7 cm. long and 1.6 cm. in dia. Two of the pieces show no signs of working except for the burred ends; the last is cut off at the far end. Most of the waste pieces were recovered from Units IV and V (see Table 16).

Woodworking residue was found at Juke Box Cave (Jennings 1957:193), Hogup Cave (Dalley 1970:181), various caves in New Mexico reported by Cosgrove (1947, Fig. 137), and Sand Dune Cave (Lindsay 1968: 71, Fig. 45). Burred sticks were found in all of these sites except Sand Dune Cave.

Wooden Hoop

A twig of *Rhus trilobata* 0.6 cm. in dia. was found bent in a circle and bound in two places with sections of bark of the same material (see Fig. 34b). The hoop measures roughly 13 cm. in dia. and comes from Stratum IVc.

Wooden hoops of this type are reported by Cosgrove (1947, Figs. 103 and 111) who thinks they may be portions of ring baskets, and Lindsay (1968:116, Fig. 93).

Bent Stick

A split twig from Stratum Va was found bent into a U-shape with one leg longer than the other (see Fig. 34h). Its total length is ca. 22.5 cm. and the thickness is ca. 0.7 cm. It is not noticeably worked or smoothed.

Bound Twigs

Four twigs or split twigs (see Fig. 36w-z) were found which are wound loosely with bark strips: three are wound with finer vegetal fiber, one is merely knotted back on itself; one has a short cord wound around it. They are not formally distinctive nor do any of them show what could be considered special attention or care in manufacture. All the twigs are under 17 cm. in length and average ca. 4 cm. in dia. They were distributed by provenience as follows: 1--surface, 1--Vc, 3--Vb, 1--IIIg, 1--TT4, 2--NP (see Table 17).

Cosgrove (1947, Fig. 118) reports bound twigs with some similarities to the Cowboy Cave artifacts. He conjectures that they may be *pahos*. I would rather believe that these twigs represent some non-productive activity which could be compared to doodling.

Split Wood

Three split pieces of wood were found at Cowboy Cave which were not distinctive; they are rather thin and flat. They measure 13 cm. long, 0.7 cm. wide, and 0.2 cm. thick; 25 cm. long, 1.8 cm. wide, and 0.3 cm. thick; 10 cm. long, 1 cm. wide, and 0.3 cm. thick; and come from Strata IIIe, Vd, and IVb, respectively.

Miscellaneous Worked Wood

Twenty-four sticks are either too fragmentary or not distinctive enough to assign to a category other than miscellaneous. Fifteen of these are shafts of varying lengths, six are split sticks with some wear or cutting on the ends, and three are charred points. See the provenience chart (Table 16) for their distribution.

Decorated Reed

Three short sections of *Phragmites* reed were found which had been carefully split lengthwise and shaped and decorated (see Fig. 36l-n). One, from Stratum Vc, measures 2.1 cm. long and 0.9 cm. in dia. and is cut squarely across both ends. One end is lightly burned for a length of 0.4 cm. and the corners are cut to make it somewhat wedge-shaped. Both of the remaining split sections have been incised across the exterior convex curve of the reed; one has six marks filled with charcoal, and the other has two marks which are filled with dirt only. The curved interior of both sections has been partially coated with pitch and, in the case of the piece with six incisions, rubbed with charcoal or some other black material. The section with two marks measures 2.4 cm. long and 0.9 cm. in dia. and has no provenience, while the other, with six marks, measures 3 cm. long and 0.9 cm. in dia. and comes from Stratum Vc.

Incised split reed sections quite similar to these from Cowboy Cave were found in Triangle Cave in Harris Wash near Escalante, Utah (Fowler 1963:63, Fig. 29), while split reeds with some resemblances were found at Gypsum Cave, Nevada (Harrington 1933:146, Fig. 57a, b). They are speculated to be gaming pieces or possibly broken reed tube beads. The finished edges of the Cowboy Cave artifacts suggest that they were used as gaming pieces or counters.

Reed Sections

Four short reed sections were found; all contain one joint. Three of them (see Fig. 36o-q) are cut off carefully at both ends. The longest of these measures 5.2 cm. long, 1 cm. in dia. and has no provenience. Another, from Stratum Vc, is 2.1 cm. long and 0.9 cm. in dia. and appears to have been hollowed out to the joint from both ends. The third section, also from Stratum Vc, and which may also have been hollowed out, measures 2 cm. in length and 0.85 cm. in dia. The last reed section is broken and one end is cut off at an angle. It measures 1.85 cm. in total length and 1 cm.

in dia. at the cut end, and has no provenience.

Dalley (1970:168) notes that the single jointed reed sections from Hogup Cave are similar to reed cigarettes described by Cosgrove (1947:121-122) except that the Hogup specimens, like the Cowboy Cave sections, lack scorching or piercing of the septum, which leaves their use in doubt.

Reeds with Fiber Knobs

Among the reed artifacts found at Cowboy Cave are eight sections of *Phragmites* which are wrapped at one end with either fibrous bark or thin strips of split wood (Fig. 36a-d). The wrapping spirals up the reed for about 8 cm. and is then wound around the end of the reed several times to form a knob 1 to 1.5 cm. in dia. The longest of these reeds--the only specimen that is cut off rather than broken at the unwrapped end--is 27.5 cm. The other specimens vary in length from 2.7 to 16 cm. A divergent specimen (Fig. 36d) does not have the wrapping spiraling up the reed and has not been wrapped to form a knob. Instead, a flat strip of bark has been wrapped around itself some 9 or 10 times to form a 0.7 cm. thick, flat, bark disk about 2 cm. in dia. The distribution of these artifacts by strata are: 3--IVc, 2--IVd, 1--Va, 1--Vb, 1--Vc.

The only artifacts noted which resemble these reeds with fiber knobs are from the Davis Kiva Site in Davis Gulch which flows into the Escalante River about 8 km. from its juncture with the Colorado--about 150 km. south of Cowboy Cave (Gunnerson 1959:141, Fig. 42). Gunnerson calls them problematical objects.

Worked Reed

Nineteen pieces of *Phragmites* reed were determined to be cultural but were too fragmentary and indistinct to be identified functionally. All but two of these are whole or split sections of reed which have been cut off or scored at some place along their length. The others are slender (0.35 cm. in dia.), uncut sections which have been wrapped

with wood bark fibers. One of these has been wrapped only five or six times near one end, while the other is wrapped several times at one end and has considerable wrapping at the other as well. These wrapped specimens are from Strata IIIh and IIIf, while the others are distributed by strata as follows: 4--NP, 2--IVc, 1--IVd, 2--Va, 8--Vb.

Split-twig Figurines

Twenty-two zoomorphic split-twig figurines were found at Cowboy Cave. Of these, nine are considered to be complete, while the remainder are incomplete or unfinished. These figurines are adroitly made by splitting a length of *Salix* or *Rhus trilobata*, bending it, and wrapping it back on itself several times to create a four-footed animal, complete, in some instances, with head, ears, and muzzle. In general, these figurines have back legs longer than the front, a rather elongated neck, and vertical as well as horizontal body wrapping.

The largest of the complete figurines (Fig. 37g) is also the best made and conforms to the description above, except that the neck is quite short, measuring only 1 cm. long. Overall, this figurine measures 10 cm. from the tip of the nose to the tail, 9.5 cm. from the top of the ears to the bottom of the back legs, and 2 cm. thick in the body. Unfortunately, this figurine has no provenience.

Another well-made figurine (Fig. 37d) is from Stratum IVc and measures 3.5 cm. from rear to nose, 8.8 cm. from ears to toes, and 1 cm. thick. This figure has the elongated neck, and back legs considerably longer than the front. It is wrapped around the neck and front legs with grass.

A complete figurine from Stratum IVd also shows the elongated neck and back legs, although the front legs have been charred and possibly burned off for an undetermined length. This figurine (Fig. 37h) measures 5.7 cm. back to front, 10 cm. from top to bottom, and 2 cm. in thickness.

Another complete figurine (Fig. 37a) with elongated neck and legs and a rather

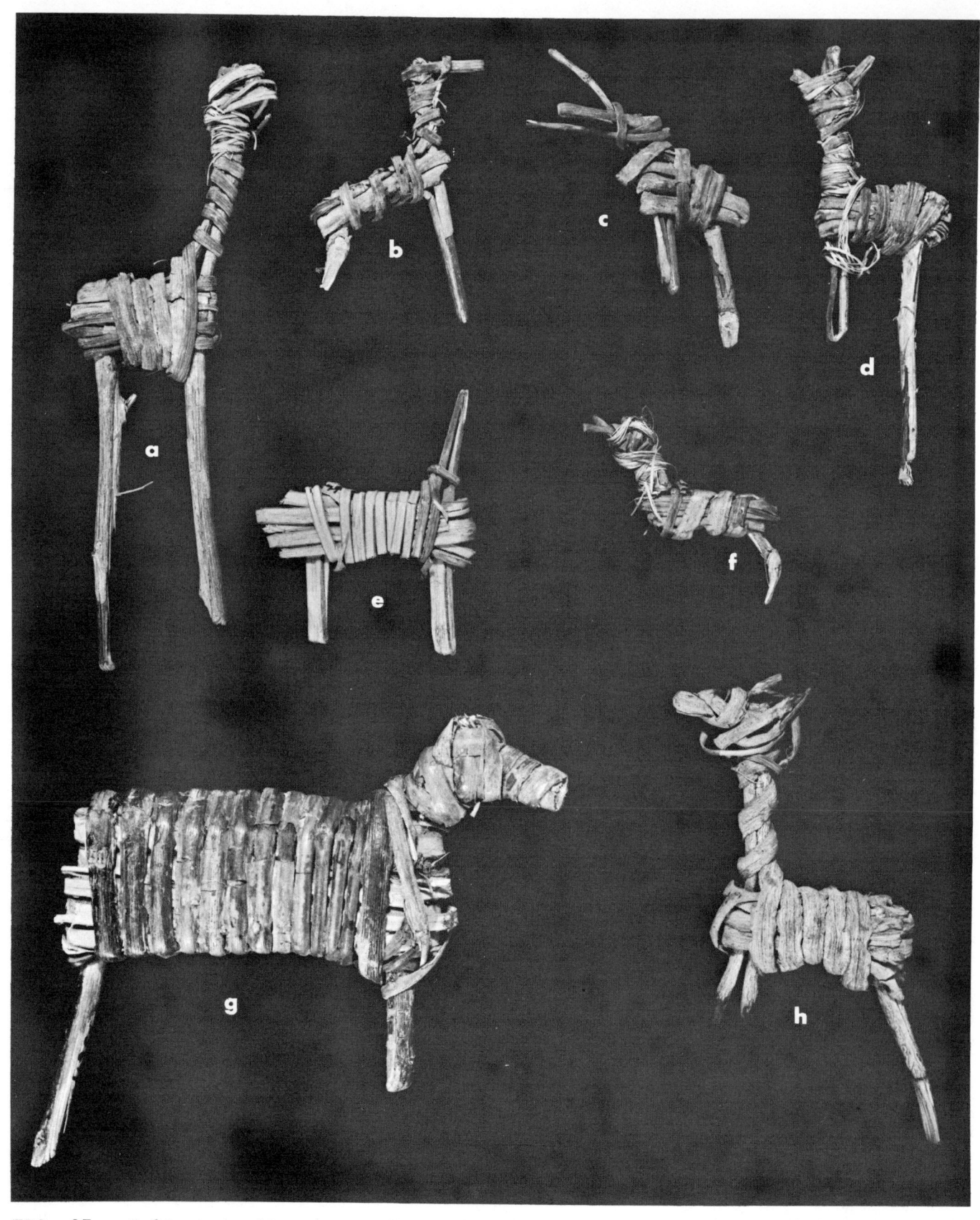

FIG. 37. Split-twig figurines. Proveniences: a, IVc; b, IVc; c, NP; d, IVc; e, Vb; f, IVd; g, NP; h, IVd.

knobby, formless head wrapped several times with grassy fibers was recovered from Stratum IVc. It measures 4.5 cm. front to back, 13 cm. top to bottom, and 1.5 cm. thick. A cruder and smaller, though apparently complete figure (Fig. 37b) also found in Stratum IVc has front legs longer than the back ones, and a long neck. This figurine, like another illustrated (Fig. 37f), has as back legs the unaltered stub where the twig was attached to a larger branch. The former specimen measures 4 cm. front to back, 5.5 cm. top to bottom, and 1 cm. thick. The latter is small, rather poorly made, and lacks front legs. This artifact (Fig. 37f) from Stratum IVd, measures 4 cm. front to back, 4.5 cm. top to bottom, and 0.85 cm. thick. The head and neck region are wrapped with grass.

The crudest of the complete figurines is rather haphazardly wrapped in the body and neck regions and has two very slender twigs emerging from the head area. This figurine (Fig. 37c) measures ca. 4 cm. front to back, 7 cm. top to bottom, and 1.2 cm. in thickness. It has no provenience.

The last complete figurine (Fig. 37e) is constructionally divergent in that the back and front legs, as well as the neck, are all made of split twigs doubled back on themselves, the neck is not wrapped, most of the body portion is wrapped with split wood without bark, and there is no representation of a head. This figure comes from Stratum Vb and measures 4.5 cm. front to back, 5.5 cm. top to bottom, and 1.2 cm. thick.

In addition to the figurines already described, there were two figurines (Fig. 38) that were lost when they were sent to be radiocarbon dated. They were photographed in the University of Utah archeology laboratory prior to being sent, so some descriptive comments can be made, although they were not examined for this analysis. One was apparently complete, while the front legs were missing on the other. They were very similar to Figure 37b and f in that both were small, rather crudely made, and their back legs formed from unaltered stubs of a larger branch. From the scale on the photograph they were 8.7 cm. high by 9.2 cm. long, for the complete figure, and 6.5 cm. high by 9.2 cm. long for the broken one. Both came from Stratum IVd.

The remaining figurines are either broken or unfinished and vary little from the descriptions above. Two of these consist of sticks which have been split and have some wrapping but are not complete enough to allow then to be positively identified as figurines (Fig. 39a).

FIGURINE CONSTRUCTION. In order to understand better the constructional details of the figurines it was decided to take two of them apart. Of the two chosen, one (Fig. 37h) was selected as being typical of the type encountered at Cowboy Cave, while the other (Fig. 37e) appeared to be a divergent form and was examined to determine the extent of that divergence.

Each figurine was soaked for a short time in distilled water and then carefully unraveled while an artist made step-by-step drawings of the construction process in reverse. Figure 40 is a reproduction of these drawings (A^1-I^1), as well as a comparison of the Green River figurine with the typical Grand Canyon figurine (A-I) as described by Wheeler (1942). The divergent Cowboy Cave figurine is not illustrated.

The excellent state of preservation of the twigs, coupled with the soaking process, enabled us not only to take the figurines apart but, in the case of the typical form, to reconstruct it as well. Each of these figurines was made from a single, unbroken length of a split twig, although there were differences in construction.

The first step (Fig. $40A^1$, B^1) in the construction of the typical figure consisted of splitting an unaltered green twig from the small distal end back toward the proximal end, leaving a short section unsplit. Leaving this solid section to represent the back legs, one of the twig halves was bent at a right angle to form the front legs (Fig. $40C^1$). The twig half was then doubled back on itself again to form the foundation for the head and neck region. It was then bent at a

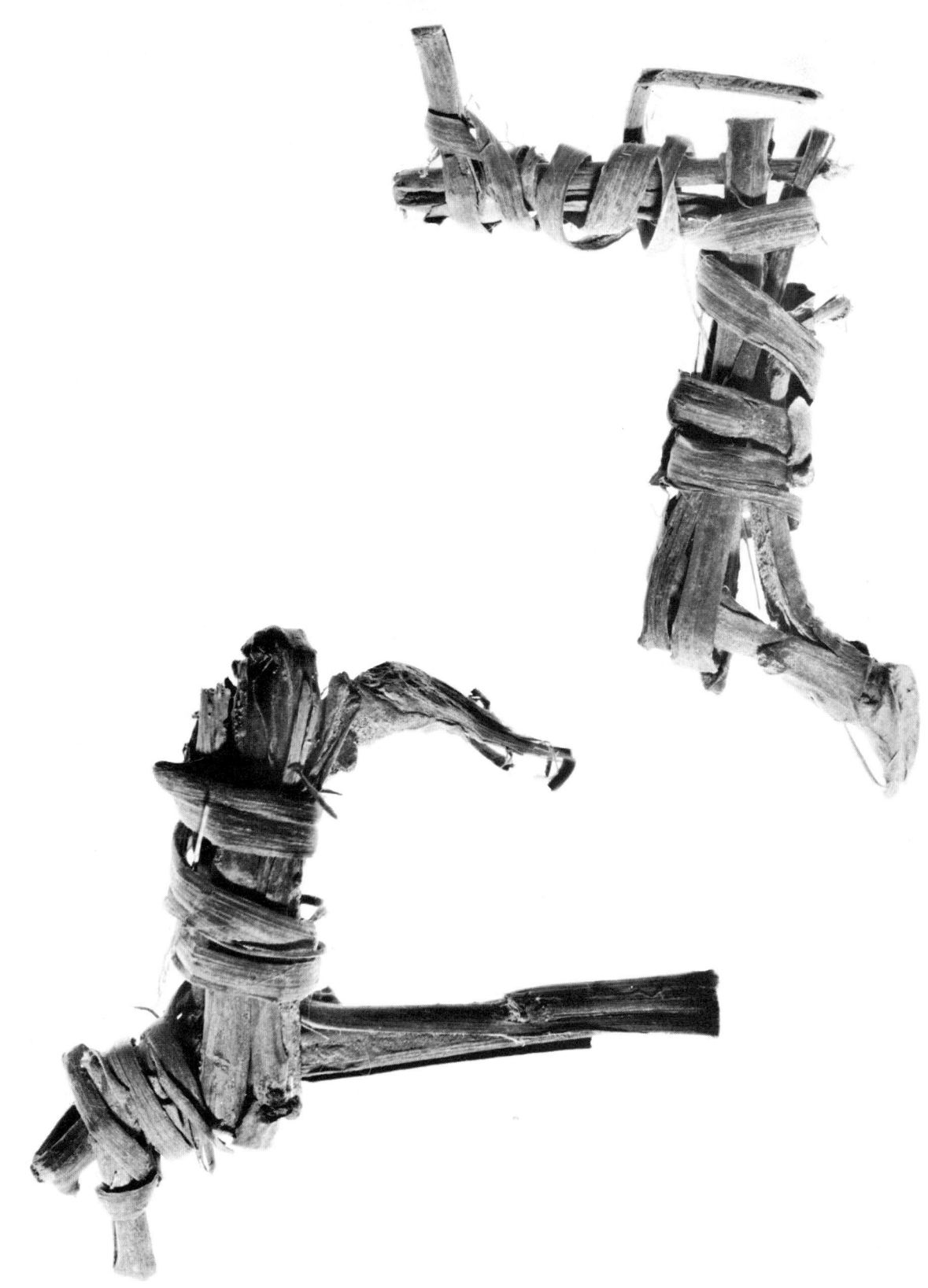

FIG. 38. Lost figurines from IVd.

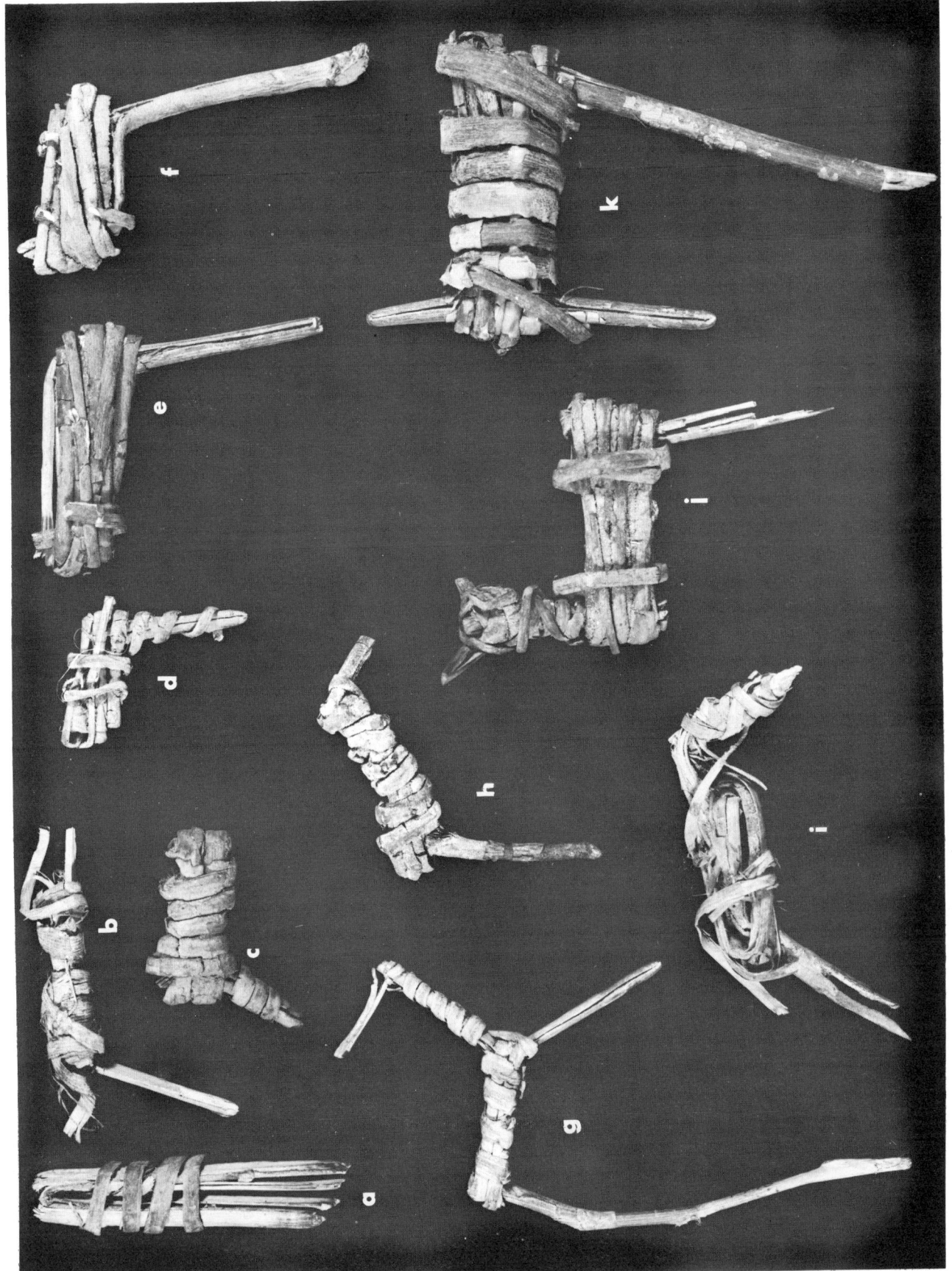

FIG. 39. Broken figurines. Proveniences: a, Va; b, NP; c, Vb; d, IVc; e, Va; f, Vb; g, IVd; h, NP; i, IVc; j, Va; k, NP.

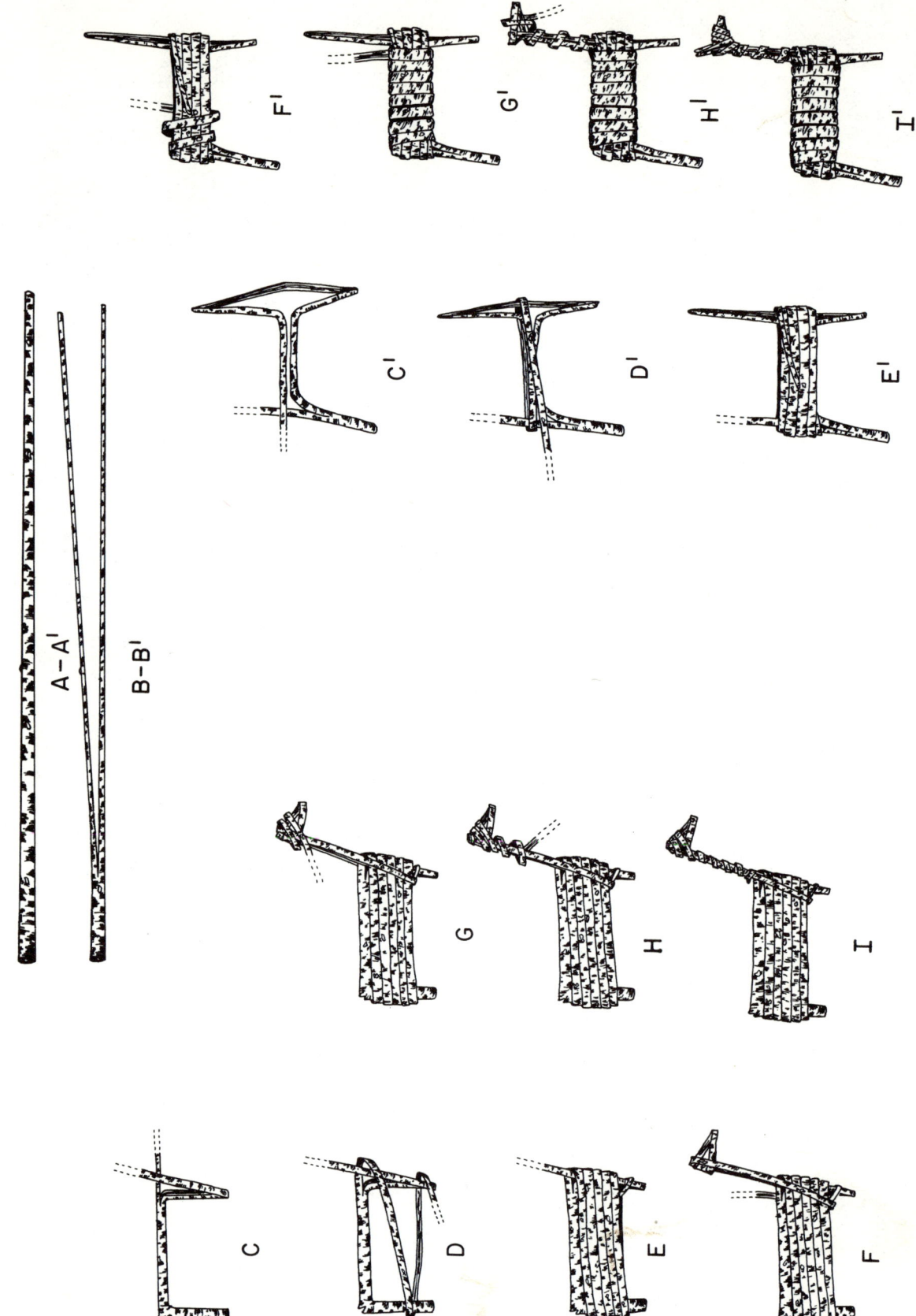

FIG. 40. Constructional sequence for split-twig figurines from the Grand Canyon area (A-I after Wheeler 1942) and the Green River area (A'-I').

right angle once more to lend additional support to the body. After this, the remaining length of this twig was wrapped horizontally around the front and back legs to fill out the body (Fig. 40D[1]). The loose end was tucked under one of these body wrappings (Fig. 40E[1]).

The other twig half was not used to finish the body by wrapping it several times vertically from the back toward the front, a process which also served to hold the tucked end more securely (Fig. 40F[1], G[1]). This twig half was then spiraled up the neck and the head, muzzle, and ears were constructed by extending the twig for a short distance, binding it back on itself and securing it with a wrap (Fig. 40H[1]). This was done for the muzzle first, then each ear was formed in turn, after which the loose end was tucked into one of the spirals in the neck region.

The divergent figurine differed initially in that it was made from a twig split into thirds for its entire length, with the middle third serving as the material for this figurine. Additional differences occur in the starting point and the formation of the back legs. The bending began some distance above the larger proximal end where the split twig was bent and doubled back on itself to form the rear legs, extended to form part of the body, then bent back to form the front legs, then up to form the neck. The larger end was then wrapped horizontally to form a portion of the body, then tucked in. The distal end was also wrapped horizontally, but then it was wrapped vertically around the body from back to front and, finally, looped once at the bottom of the neck region to secure the loose end.

This figurine was well-preserved, but the wood was perhaps not green at the time of its use, as it was much more brittle and was cracked and split at the bends rather than bent smoothly as was the typical figurine. As a consequence of this cracking, the figure began separating at the breaks when it was put into the distilled water, and we were unable to reconstruct it.

The observed constructional differ-ences are apparent in the twig material, the starting point, the difference in the back legs (solid twig vs. split twig), the lack of a detailed head region, and the overall finish of the two; i.e., the typical form has the bark left on the exterior, while the divergent form has the raw split wood as a finish.

Some 377 split-twig figurines have been found at 16 different sites in Utah, Arizona, Nevada, and California. All of these have been reviewed and considered by Schroedl [1976:86-110]. The occurrence of these figurines can be grouped into two geographical areas: 1) the Grand Canyon area of the Colorado and 2) the lower Green River area of Utah. This geographical differentiation is complemented by constructional and suggested functional differences, as noted below. Such a grouping ignores the California (Newberry Cave) and Nevada (Etna Cave and Moapa Valley) figurines for the moment. The limited number of samples from the latter sites, combined with the poor photos and descriptions available, make comparisons difficult, although artifactual similarities, e.g., Gypsum points, have inspired further attention to these two areas.

Constructional differences between the Grand Canyon and the Green River figurines can be seen in the illustration (Fig. 40). The initial difference is that the unsplit, proximal end of the twig forms the body core in the Grand Canyon version, but only the back legs in the Green River types. Also, the formation of the neck is generally different in the Grand Canyon type. The most obvious difference in the finished figurine is that the Grand Canyon body consists of horizontal wrappings only, while the Green River type is finished with an additional series of vertical wrappings.

Generally speaking, there seems to be more morphological and constructional variation in the Green River figurines, e.g., Moab (Pierson and Anderson 1975: 44), Cottonwood Cave (Gunnerson 1969), Moonshine Cave (Pendergast 1961), Green River (Tripp 1967), and Walters Cave (Jennings 1975). Considerable variation in form can be seen even among the complete and broken figurines from Cowboy Cave. In the Grand Canyon, however, the forms are more consistent, as seen in examples from Stantons Cave (Jett 1968), Sycamore Canyon (Kelly 1966), Tse-An-Kaetan (Schwartz et al. 1958), and others.

This morphological consistency in the Grand Canyon area is interesting in light of the functional change suggested by Schroedl (1976:109). The Grand Canyon figures are reported from caches in cave walls or from surface scatters, and have been assigned a magico-religious role because of the places they have been found and the association of wooden "spears" piercing their bodies. The Green River figurines, on the other hand, are not found in caches, but occur scattered within cultural deposits. Nor have any cases of "speared" figurines been reported from the Green River area. On these bases Schroedl hypothesizes a functional change from a magico-religious hunting aid to a less formal, more mundane role. The former might demand more consistent formal representations of the animals hunted, e.g., deer (Reilly 1966:135) or mountain sheep (Jett 1968:347-8); whereas, in the latter situation, a wider variety of forms could be expected to occur.

Split-twig figurines are firmly associated with the late Archaic in the Green River region, where they persist as the transition into Basketmaker times takes place. On the basis of radiocarbon dates, Schroedl [1976:109] says that the more functionally specific figurines of the Grand Canyon appeared earlier than the Green River forms. However, radiocarbon dates from Cowboy Cave place most of the figurines at ca. 3500 B.P., making them more or less coeval with the Grand Canyon group. Hopefully, figurines with diagnostic cultural affiliations will eventually be found in the Grand Canyon region.

SUMMARY

In theory, the dry environment and the well-documented stages of occupation at Cowboy Cave should enable prehistorians to consider change in the technology associated with the wooden artifacts. Here, on the contrary, one is struck by the apparent *continuity* in woodworking techniques through time. This technological persistence has a broad areal base, as similar tools and woodworking methods are found in sites ranging from Nevada, Utah, and Arizona to New Mexico and west Texas.

The only major technological innovation at Cowboy Cave occurs during the transition from Unit IV to Unit V, and that is the appearance of arrow parts and, by implication, the bow. Although evidence is thin, several arrow mainshafts and foreshafts do appear in Unit V, along with Rose Springs projectile points. Atlatl shafts and Gypsum points persist after the transition, although in slightly decreased numbers. The appearance of the bow and arrow at about this time has been noted in other Utah sites (Dalley 1970:184; Schroedl 1976: 73) and is considered a marker for the end of the Archaic and the harbinger of later agricultural groups, such as the Fremont (Schroedl 1976:73). Schroedl and Dalley both note that the appearance of the bow and arrow is the only distinctive change that occurs during the late Archaic period.

The relative abundance and variety of wood and reed artifacts suggests that they were an important component of the prehistoric tool kit in the Colorado Plateau. Activities reflected in the remains include: limited hunting, as evidenced by the few arrow parts, atlatl and dart shafts, snares, and bone; seed harvesting, as evidenced by the digging stick with seeds imbedded and the various sticks with pitch which were probably used in the manufacture or repair of seed gathering basketry; and social or leisure activities, suggested by the occurrence of wood and reed gaming counters, drilled items, bound twigs, and split-twig figurines.

Although only limited identification was made on the material used, some preferences are apparent. The more straight-grained woods, such as *Cercocarpus* and

Populus, were preferred for dart, arrow,
and other shafts, while the tough, fibrous chenopods were used for tools
which required twisting or bending.
This latter group includes the wooden
awl, all of the scissor snares, and the
spindle whorl shafts.

Less common wood artifacts from Cowboy Cave include wooden gaming pieces
and drilled items, noose snares, reeds
with fiber knots, and a divergent atlatl
form. The figurines, which have already
been discussed at some length, are also
of considerable interest, as they are
comparable to figures from the Grand
Canyon region, Nevada, and even California. Associated dates from Cowboy
Cave place the split-twig figurines in
about the same period as they appear to
have been popular elsewhere, though they
persist at Cowboy some 1,500 years
longer.

BONE AND SHELL MATERIAL

William A. Lucius

Osteological analysis of material from Cowboy and Walters Caves was undertaken on three classes of bone remains: bone artifacts, human bone, and unmodified animal bone. Comparative collections of the University of Utah Archeological Laboratory and the Utah Museum of Natural History provided the bases for identifications. In addition, Gilbert (1973) and Lawrence (1951) served as sourcebooks for bone identifications. John Wycoff of the Utah Museum of Natural History provided assistance with avifaunal remains and problem identifications.

BONE AND SHELL ARTIFACTS

During initial separation of the osteological remains, any bone exhibiting alteration from its usual form was considered an artifact of human activity. Generally, these fell into three subclasses: tools, ornaments, and miscellaneous objects (Tables 17, 18).

Tools

Many artifacts were recognized that fit the generally accepted description of awls and flakers (Haury 1975:375-379), having either pointed ends, in the case of awls, or blunt ends, in the case of flakers (Fig. 41a, b). Specific identification of the parent bones used in their manufacture was impossible as in every case the diagnostic features were missing. However, most of the tools were manufactured from splinters of artiodactyl long bones (Fig. 41). The remainder were either modified artiodactyl scapula fragments or portions of long bones fron unidentified smaller mammals.

An awl bundle may be a basketmaking tool kit. It consists of a splinter bone awl with a sharpened end bound together with two unmodified splinters of bone. These are wrapped with a grass pad and secured by a yucca tie (Fig. 41). Most of the tools that are earlier than Unit IV were made from small mammal bones, while those in Units IV and V are predominantly artiodactyl bone fragments. This supports the results of the identifiable bone analysis in that large mammals do not appear in the record until Unit IV (Tables 19, 20)

Ornaments

Ornaments, the second class of artifacts, were divided into subclasses of beads and pendants (Tables 19, 20). As all shell artifacts recovered from the sites fall into the same categories, they are included in the discussion. A quantity of variously shaped beads were recovered from the deposits, including a spire-lopped *Olivella* shell bead (Fig. 42n) and a circular, dome-shaped, centrally drilled shell bead. Most bead forms of bone can be characterized as tubular (Fig. 42m), or shaped tubular

TABLE 17

DISTRIBUTION OF BONE AND SHELL ARTIFACTS, COWBOY CAVE

	NP	Test Trench				I		II		III											IV				V				Sur	TOTAL
Stratum:		1	2	3	4	a	b	a	b	a	b	c	d	e	f	g	h	i	j	k	a	b	c	d	a	b	c	d		
Tools (Awls)																														
Pointed	7	1				1	1			1					2			1					5	2		2	4	1	2	30
Blunt																1									2	2				5
Ornaments																														
Beads	3												1				1							1	2	15		1	1	25
Pendants	1												1															1	1	4
Gaming Pieces													1													3				4
Flattened Bones	2																						1	1		2	1		2	9
Blanks	1															1									1					3
Tubes	3													1									1							5
Phalanges															2	1			1											4
Debris	1								1				2				1		1			1	3	1	1	2	1	1	1	17
Fishhook																										1				1
Shell									1								1								1				1	4
Unknown	1									2												1	2			2			3	11
TOTAL	19	1				1	1		2	3			5	1	4	3	3	1	2			2	12	5	7	29	6	4	11	122

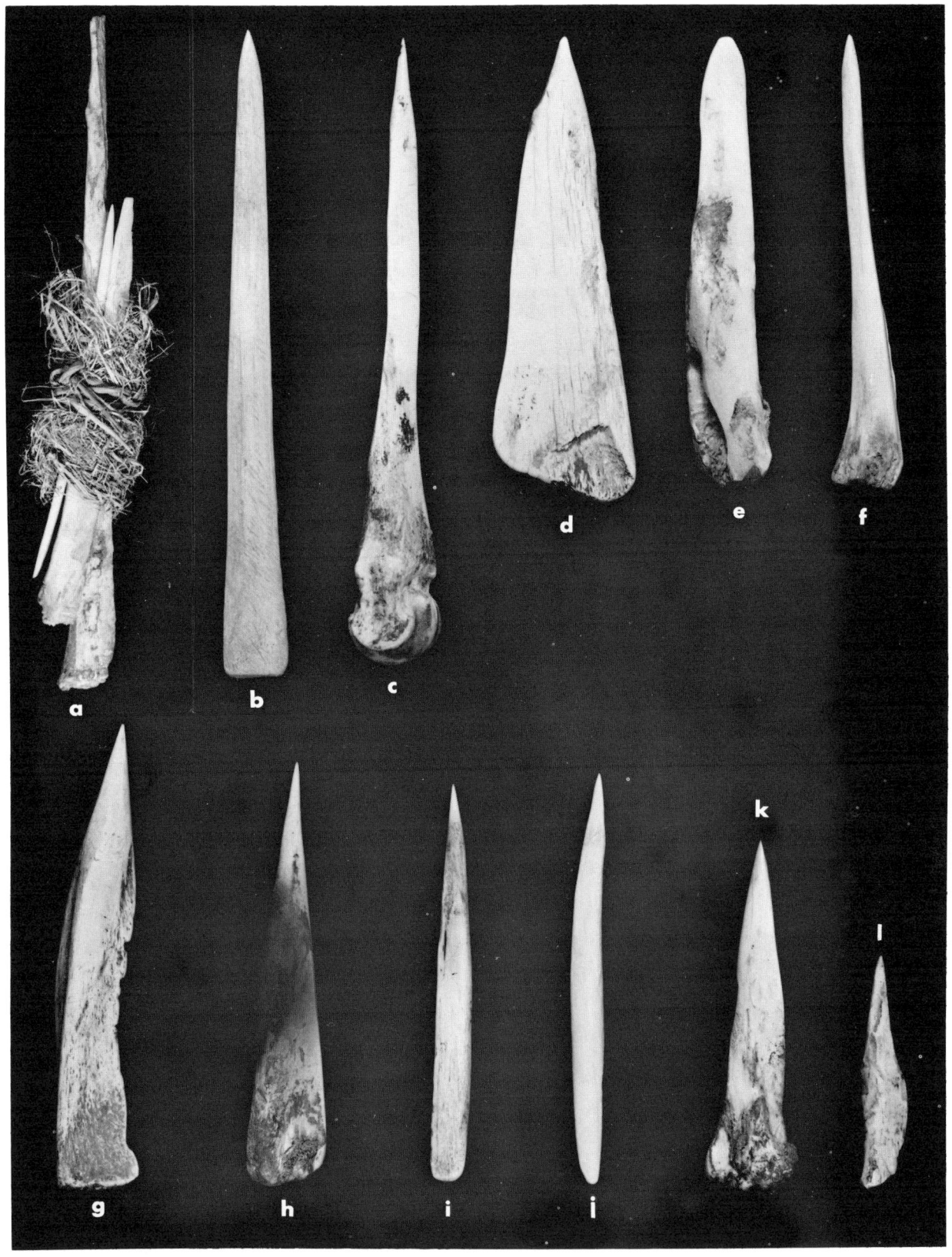

FIG. 41. Bone tools: a, awl bundle wrapped with grass pad and yucca tie; b–1,
awls (notice range of finish and size); e, h, and k retain tendons. b measures
13.5 cm. in length.

TABLE 18

DISTRIBUTION OF BONE ARTIFACTS, WALTERS CAVE

		Unit:	I		II	III				IV		V			
NP	Test Trench	Stratum:	A	B	A	A	B	C	D	A	B	A	B	Sur	TOTAL
Pointed Tools (Awls) 1									1				1		3
Pendants													7		7
Gaming Pieces 1															1
Debris												2	1		3
TOTAL 2									1			2	9		14

(Fig. 42p). Two short tubular bead starts were apparently broken during manufacture, evidence that beadmaking occurred at the site.

A number of hyoid pendants, both drilled and unperforated, were recovered in both caves (Fig. 42d-k), as were two pendants of unidentified shell. In addition, a deer-hoof tinkler similar to those figured in Lindsay et al. (1968: 57, Fig. 36) was recovered (Fig. 42u). The pendants appear to be the remains of an individual hyoid bone necklace from Walters Cave. Most of the bones retain sinew and leather wrappings, and their bleached appearance may indicate they had been long exposed to sunlight, an uncommon characteristic of most bones in the deposits. Both drilled and unperforated pendants are present on the necklace (Fig. 42e). Similar necklaces are present in private collections; their scarcity in Archaic collections may be in part due to the difficulty of distinguishing those items as artifacts unless they retain their sinew/leather attachments (Lindsay et al. 1968:58, Fig. 37). A number of possible hyoid pendants were recovered from the deposits, but the lack of perforations or

attachments prevented their positive identification as pendants.

A portion of a necklace or possibly a wristlet with bird talons threaded on a leather thong (Fig. 42l) is of uncertain provenience. The talons were evidently from a falcon or other raptor not identified.

Miscellaneous Artifacts

The final category of worked bone has a number of subclasses, each containing only a few items. Five flat, smoothed, bone tablets, two with a distinct ovoid shape and incised decoration on one side and a roughened surface on the other (Fig. 42b, c), fall within the description of gaming pieces (Lindsay et al. 1968:58-59). With two exceptions, the gaming pieces are from Stratum Vb.

One object resembles a composite fishhook. It is also from Stratum Vb, and is of an unusual design not previously reported in the literature. Two heat-tempered bone points are bound to a bone shaft with cord. There is no evidence of attachment to a line (Fig. 42r).

Four broken rabbit phalanges with grass stems placed in their marrow cavities were found in Unit III; their

TABLE 19

MINIMUM COUNT OF IDENTIFIABLE MAMMAL SPECIES AT COWBOY CAVE

Species	NP	TT 1	TT 2	TT 3	TT 4	I a	I b	II a	II b	III a	III b	III c	III d	III e	III f	III g	III h	III i	III j	III k	IV a	IV b	IV c	IV d	V a	V b	V c	V d	Sur	TOTAL
Cottontail	4	1			1	1*	1*	2					2	1	1	1		1		1	2	1*	1	1	2	5	1	1	1*	32
Jackrabbit	2	1							1				1	1	3	1	1*	1			2		1	1	2	2			1*	21
Mule Deer	1*																	1					1			1			1*	5
Mountain Sheep	2		1*																		1		1		1	1	2		1	10
Prairie Dog	1*																								1	3	1*	1*		7
Porcupine															1*															1
Canis	1*					1*		1*																						3
Bison								1*																						1
Elk																											1*			1
Lynx	1*																													1
Human																									1*					1
TOTAL	12	2	1		1	2	1	4	1				3	2	5	2	1	3		1	5	1	4	2	7	12	5	2	4	83

*Indicates minimum count derived from one element.

TABLE 20

MINIMUM COUNT OF IDENTIFIABLE MAMMAL SPECIES
AT WALTERS CAVE

| | NP | TT | Unit: I | | II | III | | | | IV | | V | | Sur | TOTAL |
			Stratum: A	B	A	A	B	C	D	A	B	A	B		
Cottontail	2	1*							1	1	1*	1			7
Jackrabbit									1	1	1		1*		4
Badger									1*						1
Human	1														1
TOTAL	3	1							3	2	2	1	1		13

*Indicates minimum count derived from one element.

significance is unknown.

Bone splinters with one or two flattened faces were found to be restricted to Units IV and V. Their fragmentary nature precludes any statement of their use. Additionally, several bone blanks of unknown utility were recovered. Of the several bone tubes in the collection, one large, charred item was well smoothed and exhibits central roughening, perhaps for the attachment of cordage or sinew (Fig. 42a).

Manufacturing debris, patently bone debris left from the manufacture of beads and tubes, was found throughout all strata, adding to the evidence that tool and ornament manufacturing occurred at the sites.

In an unknown category were placed other items of bone, including ochre stained fragments, and a bone with adhering organic material. There were a few fragments of bone which appear to have been modified but take no identifiable form. In general the numbers and varieties of bone artifacts are greatest in Units IV and V, as was the case with the identifiable bone remains and raw meat weights (Tables 19, 20, 21, 22).

HUMAN BONE

There were two occurrences of human bone, one in each cave. A fragmentary eroded maxilla with four well-worn teeth was found in Cowboy Cave, Stratum Va. The second occurrence of bone was in a disturbed burial just within the dripline of Walters Cave (Tables 19, 20). As the remains were noted only during the backfilling of the cave, no provenience could be assigned. Only phalanges, several vertebrae bodies, and numerous long bone fragments were found. Because of the incomplete nature of the skeletal remains, no sex or age ascription was possible.

UNMODIFIED ANIMAL BONE

Unmodified animal bone, which comprises the bulk of the osseous remains, was initially sorted into identifiable and nonidentifiable categories. Those bones assignable only to an order level were usually considered unidentifiable. Derivation of minimum counts and recording of data followed the procedure outlined in Sudden Shelter (Jennings et al.

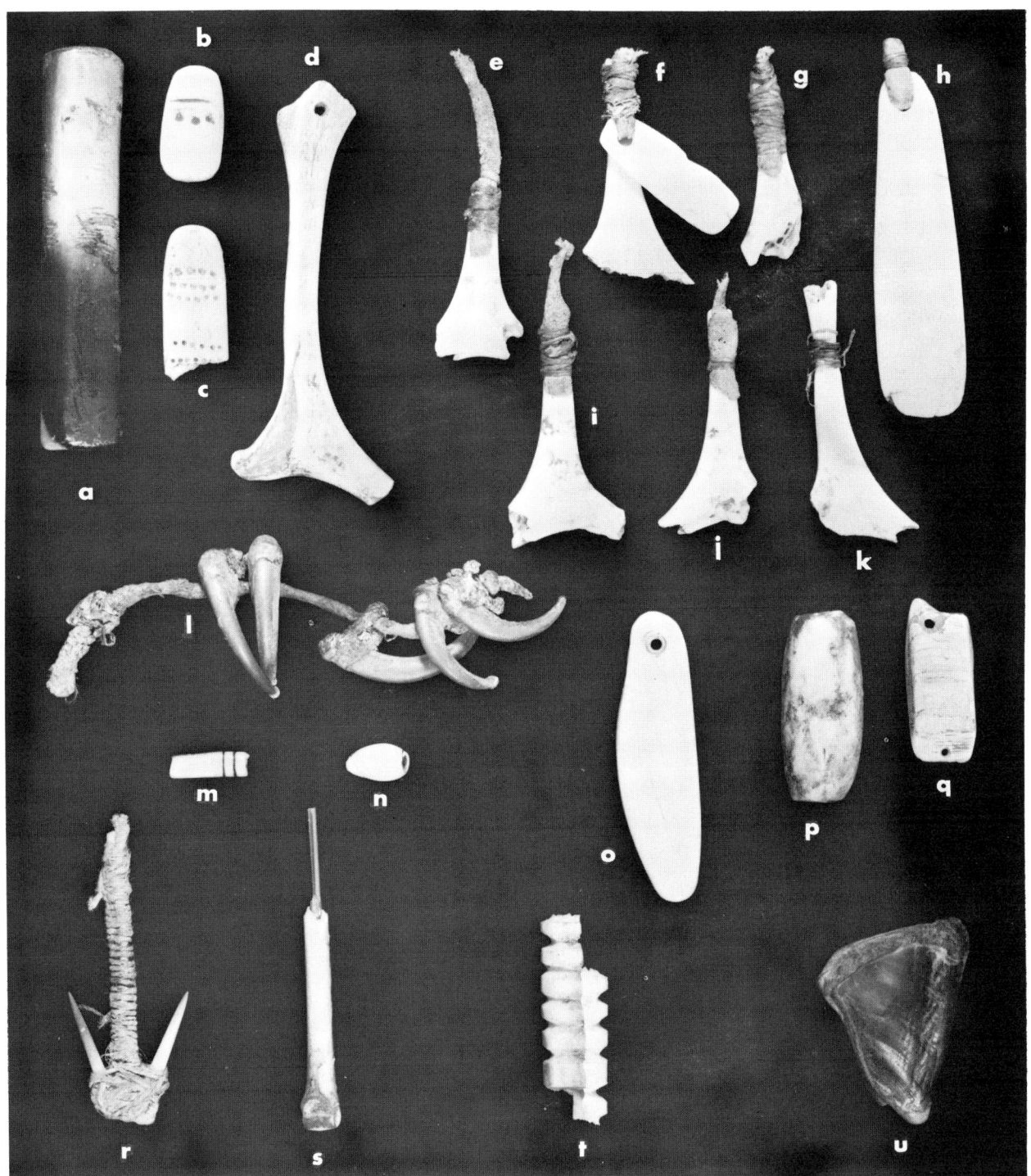

FIG. 42. Miscellaneous bone and shell: a, bone tube with central roughening (5.8 cm. in length); b, c, bone gaming pieces; d, artiodactyl perforated hyoid pendant; e-k, artiodactyl hyoid bone pendants with sinew and leather attachments; l, prairie falcon (?) claw necklace on leather cord; m, tubular bone bead with incised decoration; n, spire-lopped *olivella* shell bead; o, perforated bone pendant; p, tubular bone bead; q, perforated shell ornament; r, composite bone fishhook wrapped with cord; s, jackrabbit phalange with grass stem inserted into cavity; t, bone beads in the manufacturing stage; u, artiodactyl hoof tinkler.

TABLE 21

MINIMUM AVAILABLE MEAT WEIGHTS FROM COWBOY CAVE

	Artiodactyla				Lagomorpha			Rodentia			Carnivora			
	Cervus canadensis	Odocoileus hemionus	Ovis canadensis	Total	Lepus californicus	Sylvilagus sp.	Total	Cynomys parvidens	Erethizon dorsatum	Total	Canis sp.	Lynx rufus	Total	TOTAL, All Species
NP		100	200	300	6	7	13	1.5		1.5	12.5	15	27.5	342
TT2			100	100	3	1.75	4.75							104.75
TT4						1.75	1.75							1.75
Ib						1.75	1.75							1.75
IIa						1.75	1.75							1.75
IIb					3	3.5	6.5	1.5		1.5				8
IIIa						1.75	1.75							1.75
IIId					3	3.5	6.5							6.5
IIIe					3	1.75	4.75							4.75
IIIf					9	1.75	10.75		10	10				20.75
IIIg					3	1.75	4.75							4.75
IIIh					3		3							3
IIIi					3	1.75	4.75				12.5		12.5	17.25
IIIj					3		3							3
IVa			100	100	6	3.5	9.5							109.5
IVb						1.75	1.75							1.75
IVc			100	100	3	1.75	4.75				12.5		12.5	117.25
IVd		100		100		1.75	1.75							101.75
Va		100	100	200	6	3.5	9.5							209.5
Vb		100	100	200	6	8.75	14.75	4.5		4.5				219.25
Vc	350		200	550		1.75	1.75	1.5		1.5				553.25
Vd						1.75	1.75	1.5		1.5				3.25
Sur		100	100	200	3	1.75	4.75							204.75
TOTAL	350	500	1000	1850	63	56	119	10.5	10	20.5	37.5	15	52.5	2042

In press). Unidentifiable bone scrap was not treated further.

Thirteen species were identified in the collection, including two Lagomorpha, three Artiodactyla, three Carnivora, and five Rodentia. Aves and Reptilia elements were also distinguished, but no order or family/species levels could be assigned and, therefore, no minimum counts were derived. Of the order Rodentia, the remains of ground squirrel (*Citellus* sp.), pocket gopher (*Thomomys* sp.), and pack rat (*Neotoma* sp.) were combined in a general "rodent" category with no attempt to arrive at minimum counts. We know, however, that some portion of the rodent component was used for food by the presence of rodent bones in human coprolites; there is also ample ethnographic evidence of the use of rodents for food (Steward 1938:40).

The division of the site into cultural and noncultural strata is reflected in the bone data. With the exception of obvious rodent intrusions from other levels, which are possibly responsible for the presence of single rabbit elements in otherwise noncultural strata (Table 21), bone remains from the dung layers of deposition are characteristically stained to a yellow color (10YR7/6 on the Munsell Soil Color Chart 1975) with gray mottling. The coloration would appear to be the result of the chemical content of the dung and urine and, except for the rodent bones, all fragments from those layers are of large mammals.

The number of identifiable bone fragments from the noncultural strata of the caves is small, and only one element could be assigned a family level designation--a left mandible fragment of an apparently juvenile *Bison*. The fragment is broken along the midline and contains two cheek teeth and two emergent front teeth. The broken edge shows polishing, especially on the lower edge (Fig. 43).

Five additional bones were recovered from Unit I--two apparent tusk fragments, the articulating end of a rib, a peg-shaped tooth, and a large terminal phalange (Fig. 44).

Stratum IIb marks the initial occupation of Cowboy Cave by man, as evidenced by the occurrence of cultural bone (Lagomorpha) in some quantity (Table 17). (It should be noted that the bone recovered from Walters Cave is insufficient for analysis and, although it is listed in Table 18, the following interpretations are based on Cowboy Cave data alone). The presence of a prairie dog (*Cynomys parvidens*) manus in Stratum IIb deserves comment. With the exception of this single instance, all prairie dog remains occur in Unit V (Table 19) and consist solely of articulated forelimbs, or portions thereof. The presence of prairie dog forelimbs in Unit V may represent the debris from making skin bags, since fragments of prairie dog skin were also found.

The continuous occurrence of Lagomorpha through Unit III, to the virtual exclusion of all other animal remains, should be noted. It is not until Unit IV that artiodactyl remains appear. This suggests a shift from sole dependence on small animals to procurement of both small and large animals beginning in Unit IV.

Because the amount of unmodified animal bone present is a guide to the economic importance of the species present, conversions were made to derive available raw meat weights (Tables 21 and 22). The method of conversion used is in Jennings et al. (In press). The small amount of animal protein represented reinforces the interpretation that hunting and trapping of animals was secondary to plant and seed gathering at the sites.

SUMMARY

Evaluation of the bone analysis indicates that some shift in food procurement at the site occurred in Unit IV and continued into Unit V. The number and kind of both food animal remains and modified bones change markedly, beginning in Unit IV. The amount of food available from the animals represented is relatively small, presumably indicating that animal procurement was not the prime activity of site use.

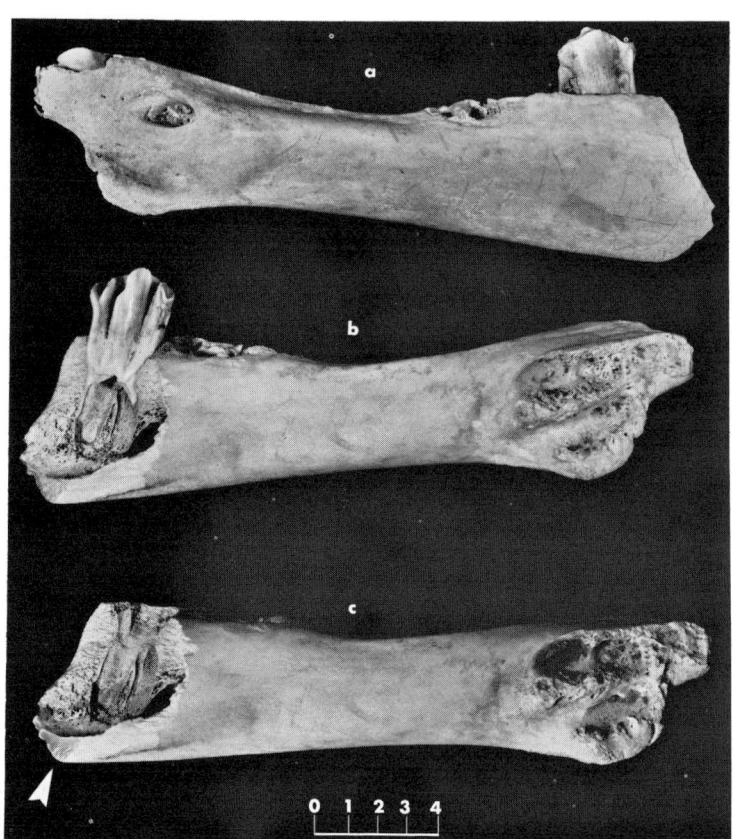

FIG. 43. Three views of juvenile bison
mandible: a, buccal view; b, lingual view
(note polishing, left end); c, same as b
with tooth missing (arrow shows wear
polish).

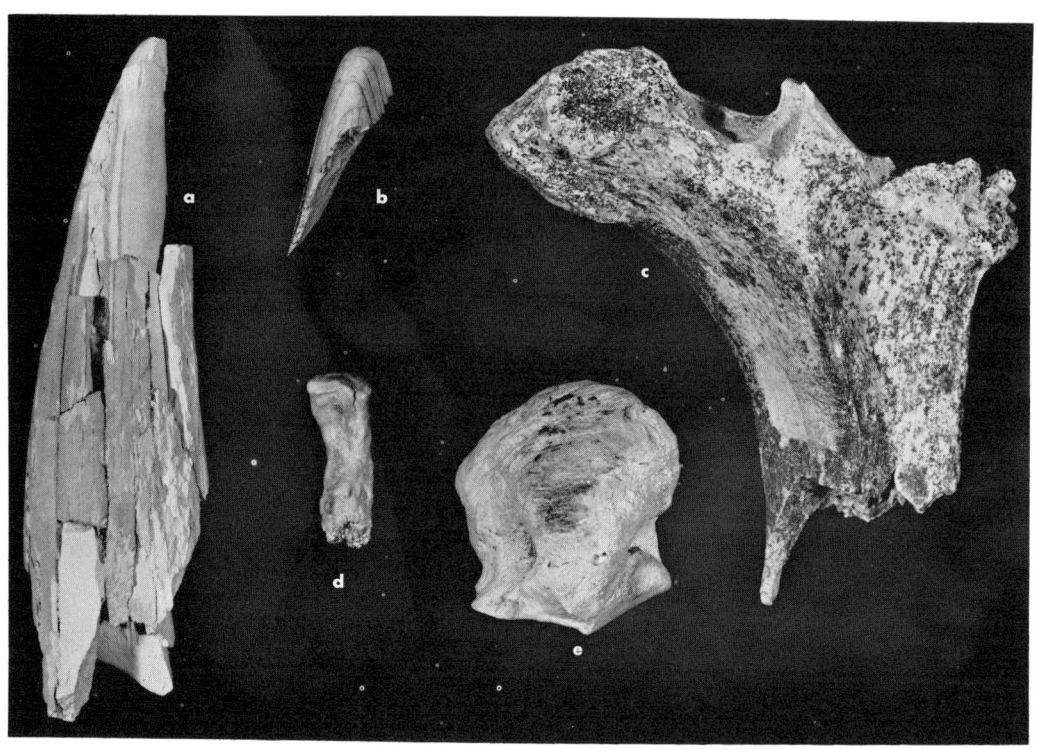

FIG. 44. Minimal bone fragments from Unit I: a,b, tusk
fragments; c, rib; d, peg tooth; e, terminal phalange.
Tusk fragment a measures 12.0 cm. in length.

TABLE 22

MINIMUM AVAILABLE MEAT WEIGHTS FROM WALTERS CAVE

	Lagomorpha	*Lepus californicus*	*Sylvilagus* sp.	Total	Carnivora	*Taxidea taxus*	Total	TOTAL, All Species
NP			2	2				2
TT			1.75	1.75				1.75
IIID		3	1.75	4.75		12.5	12.5	17.25
IVA		3	1.75	4.75				4.75
IVB		3	1.75	4.75				4.75
VA		3		3				3
VB			1.75	1.75				1.75
TOTAL		12	10.75	22.75		12.5	12.5	35.25

Comparison of the osteological remains of the sites with those of other
excavated sites of comparable age and
nature is difficult as very few of the
remains are distinctive enough to war-
rant comparison. However, the hyoid
bone pendants and the few gaming pieces
recovered (Fig. 42) show a remarkable
correspondence with artifacts from Sand
Dune Cave of northern Arizona (Lindsay
et al. 1968). The overall similarity of
the collections of Cowboy and Walters
Caves with those of the Desha Complex
have been noted elsewhere (Jennings
1975:12-15).

HIDE AND FUR ARTIFACTS

Frank W. Hull

The hide and fur collection from Cowboy Cave comprised 516 small scraps. They were most heavily concentrated in Unit V, although Units III and IV also contained substantial numbers, with maximums observed near midpoints in all levels (see Table 23). The following categories were used for convenience of description: modified soft hide, modified rawhide, strips, thongs, unmodified soft hide, unmodified rawhide, and fur.

Modified Soft Hide

The modified soft-hide category was limited to scraps of softened skin. Some pieces had cordage attached, perforations apparently for sewing, or had been sewn onto other pieces of soft hide. There are 44 pieces that ranged from fingernail-sized scraps to a piece that measured 14 cm. by 5 cm. Plant fiber cordage was attached to 12 samples and red ocher was present on three.

Modified Rawhide

Modified rawhide included all unsoftened hide scraps that showed evidence of aboriginal alteration. Bits and pieces of twisted fur strips from robes were placed in this category. The twisted fur strips, together with the 34 scraps of altered rawhide, numbered 122 pieces.

Strips

A total of 42 soft-hide strips

ranged in length from a few centimeters to one piece that was 43 cm., and in width from 0.3 to 1.9 cm. Very narrow pieces were classified as thongs. Fifteen strips were rawhide, which varied in width from 0.3 cm. to 1.5 cm. The longest piece, which had been sewn lengthwise with plant fiber cordage, measured 28 cm. in length.

Thongs

Six soft-hide thongs were recovered, the longest piece measuring 30 cm. Only two pieces of rawhide thong were found.

Unmodified Soft Hide

Unmodified soft-hide scraps were probably waste pieces from the manufacture or the repair of soft-hide articles. Small scraps were most common; the largest piece measured 13 cm. by 3.5 cm. One small scrap had red ocher on the flesh side. There were 70 scraps, of which 38 had been dehaired.

Unmodified Rawhide

The unmodified rawhide specimens were mostly small scraps that had not been dehaired. The largest of the 61 pieces measured 12 cm. by 9 cm.

Fur

There were 179 pieces of fur and tufts of hair recovered from Cowboy Cave. Techniques outlined by Trevor-Deutsch (1970) and Brown (1942) were followed in

TABLE 23

DISTRIBUTION OF HIDE AND FUR ARTIFACTS

	NP	I a	I b	II a	II b	II c	III c	III d	III e	III f	III g	III h	III i	IV a	IV b	IV c	IV d	V a	V b	V c	Sur	TOTAL
Modified Soft Hide	5				1				2	2	4	4		1		7	5	3	8	1		43
Modified Rawhide	16	2		2	5	1		1		11	7	2	2	5	3	11	10	3	29	7	3	120
Unmodified Soft Hide	6			1	1			3		4	2	4	3	5	1	7	4		11	2	3	57
Unmodified Rawhide	6										1	1		3		4	3	2	29	3	2	54
Soft-hide Strips	6						1		1	2	2		1	1		4	3	2	17	1	2	43
Rawhide Strips	2											2			1		1		6		1	13
Soft-hide Thongs	1									1		1					1		2			6
Rawhide Thongs																		1	1			2
Hair and Fur	30		1	4		1	1	11	6	13	6	6	4	3	5	17	6	8	40	7	9	178
TOTAL	72	2	1	7	7	2	2	15	9	33	22	20	10	18	10	50	33	19	143	21	20	516

an attempt to separate the hair into rodent and nonrodent groups: hairs shorter than 1.5 cm. were classified as rodent. Some larger pieces of fur were identifiable, such as woodrat (*Neotoma cinerea*), marmot (*Marmota flaviventris*), black-tailed jackrabbit (*Lepus californicus*), and Say chipmunk (*Eutamias quadrivittatus*). Several tufts of lagomorph fur seemed to be from white-tailed jackrabbit (*Lepus townsendii*) or snowshoe rabbit (*Lepus americanus*).

Numerous small tufts of rodent length hair were combined into the families Geomyidae (pocket gophers), Heteromyidae (pocket mice, kangaroo mice and kangaroo rats), and Cricetidae (native rats and mice).

Comparative collections obtained from a taxidermist provided hair samples from which some of the larger animal hair could be identified. Similarities of some of the recovered specimens required the additional use of a key to the dorsal guard hair of mammals (Moore 1974). The analysis proved that antelope (*Antilocapra americana*), mule deer (*Odocoileus hemionus*), and mountain sheep (*Ovis canadensis*) occurred from about the middle of Unit III to the surface stratum, with antelope and mountain sheep dominant.

One hair sample compared very closely with the wolf sample (*Canis lupus*). Positive identification cannot be made, however, because it is impossible to distinguish this hair from other *Canis* hair.

COMPARISONS AND CONCLUSIONS

The distribution of hide and fur (Table 23) indicates an increase through time of these animal parts in both number of specimens and kinds of artifacts produced. Although no direct correlation should be inferred between the identifiable bone and the hide and fur, similar frequencies can be seen in the occurrence of mule deer and also mountain sheep (see Tables 23 and 19). As one might expect, the distribution of hair and fur is similar to the abundance

and provenience of cottontail and jackrabbit bones. While the animals represented might not have provided large amounts of meat (see *Bone and Shell Material*), their importance as a source of raw materials for garments must not be overlooked. The frugality shown by the inhabitants of Cowboy Cave in the use and reuse of animal skins has been noted at other dry cave sites (Jennings 1957:220-223, and Aikens 1970:97-118).

GROUND STONE

James E. Dodge

There were 508 artifacts and artifact fragments in this category. They were examined, measured, and tabulated at the site where they were also briefly described. The stones were then reburied. This class of artifacts includes milling stones, manos, and grooved stones. Other worked stone is described in *Spindle Whorls, Incised and Painted Stone, and Unfired Clay Objects*.

All of the ground stone was fashioned from sandstone found in the immediate area of Cowboy Cave. Distribution in the site is summarized in Table 24. Six complete stones and 33 fragments from Walters Cave are not included in the discussion; their proveniences, however, appear in Table 25.

Milling Stones

This class comprises 35 complete specimens and 430 fragments. Sixteen of the complete stones and 81 fragments range from 5.0 cm. to 13.5 cm. in thickness and are referred to in the description below as block stones. Slab stones, which are thinner, ranging between 0.8 cm. and 4.9 cm. in thickness, comprise 19 complete specimens and 349 fragments. On all specimens where grinding motion could be determined, it was reciprocal. Nearly all (98 percent) of the milling stones were made of indurated (strongly cemented) sandstones whose cements were either siliceous or calcareous. The remaining 2 percent were fashioned from friable (weakly cemented), siliceous sandstones. Specimens which might provide useful comparison are described individually below.

BLOCK STONES. Five of these specimens are used on both sides. Fourteen are shaped by pecking and/or abrading. One complete stone, pecked and chipped to shape, is roughly rectangular. It was used on only one side. It is 49 cm. long, 27 cm. wide, and 6.1 cm. thick. Its grinding surface, which measures 33 cm. long, 15 cm. wide, and 0.4 cm. deep, is pecked and fire blackened. Material is indurated sandstone. Provenience: Vb.

A second complete specimen is unshaped, but is a roughly rectangular, tabular block. It measures 41.5 by 30.5 cm., and is 5 cm. thick. The grinding surface is 26 cm. in length, 12 cm. in width, and 1.1 cm. in depth. Material is indurated sandstone. Provenience: unknown.

A third specimen is complete and is shaped into an oval with a deep trough. The stone, of indurated sandstone, is 49 cm. by 27 cm. by 6.3 cm. The pecked trough is 41 cm. by 18 cm. by 2.8 cm. Provenience: Vc.

The proveniences of the remaining block stones appear in Table 24. No complete specimen was recovered below Stratum IVc; several fragments, however, occur in Unit III, four of which are the only specimens in this class made of

TABLE 24

DISTRIBUTION OF GROUND STONE, COWBOY CAVE

Unit:		II		III									IV				V					
Stratum:	NP	a	b	a	b	c	d	e	f	g	h	i	a	b	c	d	a	b	c	d	Sur	TOTAL
Block Milling Stones																						
Complete	5												1			1	1	3	1		4	16
Fragments	23			1		1	4					2	5		3	3	5	6		2	27	82
Slab Milling Stones																						
Complete	1	1								1	2				1	1		2	2	2	4	19
Fragments	60	6	3		7		11	7	11	6			22	2	37	20	22	31	3	7	92	349
Unmodified Manos	1								1			2	1		1		1	1			1	9
Shaped Manos	8	1								2	2		3		2	2	3	2		3	2	30
Mano Fragments	1														1	1	1	1			2	7
Grooved Stones	11																			1	5	17
TOTAL	110	8	3	1	7	1	15	7	12	11	4	4	32	2	47	28	33	46	6	15	137	529

TABLE 25

DISTRIBUTION OF GROUND STONE, WALTERS CAVE

		Unit: I		II		III				IV		V			
NP	TT	Stratum: A	B	A	B	A	B	C	D	A	B	A	B	Sur	TOTAL
Block Milling Stones															
Complete														1	1
Fragments														3	3
Slab Milling Stones															
Complete														0	0
Fragments 1						1				2		1		23	28
Manos															
Complete									1			2			3
Fragments								1				1			2
Grooved Stones												1		1	2
TOTAL 1						1		1	1	2		2		31	39

friable, siliceous sandstone.

SLAB STONES. This collection comprises 19 complete specimens and a great many fragments (349) which occurred in all occupational sequences at the cave (Table 24). Several recovered from the upper units lined cache pits (Fig. 15). Many show use on both sides and shaping by pecking/abrading.

One nearly complete, shaped slab is oval and used on both sides, both grinding surfaces having been pecked the length of the specimen. One grinding surface is slightly deeper than the other: 1.2 cm. compared to 0.5 cm. Material is sandstone. Provenience: Va.

A second complete specimen is roughly rectangular with shaping of the one use surface. It is 39 cm. in length, 32 cm. in width, and 2.7 cm. in thickness. The grinding surface has been pecked and measures 21 cm. by 20 cm. by 0.1 cm. It is indurated sandstone. Provenience: IVd.

An unshaped specimen is complete, tabular, and roughly rectangular. It is heavily stained with ocher on its single grinding surface. The specimen is 60 cm. long, 35.5 cm. wide, and 4.9 cm. thick. It has a thin, shallow grinding surface, measuring 23 cm. by 11 cm. by 0.4 cm. Material is indurated sandstone. Provenience: IVc.

The proveniences of the remaining slab milling stones appear in Table 24. All but six are indurated sandstone; of the remaining six slabs, which are friable, siliceous sandstone, one has no provenience, and the remaining five are from the surface.

Manos

Fourteen complete and 12 fragmental

manos are identifiable in the Cowboy Cave collection. All of them are of indurated sandstone. Eight complete specimens and one fragment are unmodified cobbles use-abraded on one side. They average 11.2 cm. in length, 8.7 cm. in width, and 4.5 cm. in thickness. Three come from Unit V, four from Unit III, one from the surface, and one has no provenience.

The remaining cobbles were shaped by percussion or abrading into round, oval, half-oval, or rectangular forms. All show at least one grinding surface; some have two, and two specimens have three. One specimen is ocher-stained; two show slight battering on the ends. They range in length between 8.4 and 12.7 cm., in width between 5.4 and 11.8 cm., and in thickness between 1.5 and 5.2 cm. Proveniences are shown in Table 24.

Grooved Stones

Twelve complete specimens and five fragments with one or more grooved abrading surfaces make up this artifact category. All are friable, siliceous sandstone. As Table 24 shows, all of known provenience were found either on the surface of the cave floor, or in the most recent stratum (Vd). All exhibit grooving across the top in various lengths and angles and in depth from 0.1 cm. to 0.5 cm. They range in length from 13.5 to 51 cm., in width from 7.2 to 37.4 cm., and in thickness from 3.2 to 25 cm.

SPINDLE WHORLS, INCISED AND PAINTED STONE, AND UNFIRED CLAY OBJECTS

Frank W. Hull and Nancy M. White

Various clay and stone objects were recovered from Cowboy and Walters Caves which are either man made or bear man-made decorations or markings. They are classified here in three major groups-- spindle whorls, worked and decorated stone, and unfired clay objects. Among them are clay figurines found in strata dating much earlier than the traditionally ascribed Basketmaker/Fremont affiliation.

Spindle Whorls

Twenty-three spindle whorls were recovered during the excavation of Cowboy and Walters Caves. Their sizes vary from 8.4 to 1.7 cm. in dia., with the average being 2.7 cm. Nineteen of the spindle whorls or fragments are made of unfired clay, two of pinyon pitch, and two are of sandstone. Four of the spindle whorls found have all or part of the central stick remaining, the wood being Chenopodiaceae. The sticks have been smoothed to remove any major irregularities and the distal ends worked to a point. With the exception of the stone disks, all the whorls are formed around the stick; in most of the fragments the imprint still remains.

In six clay whorls, vegetable fiber and rock inclusions were observed; they appear to have been incidental to the manufacture of the whorls. Several clay figurines recovered from the site also contain incidental inclusions.

Spindle whorls were found from Stratum IIId to the surface (Table 26). Most occurred in Strata IVd and above, and were grouped in the western third of Cowboy Cave. Five whorls from Walters Cave are included in the tabulation.

The classification of the spindle whorls follows Haury (1950). Those from Cowboy Cave fit into three of his seven categories: discoidal, ellipsoidal, and biconical. Eleven are discoidal (Fig. 45a, b). The thin variety, which includes the large pitch whorl, has nearly parallel sides. The thick discoidal whorls have a convex cross section. Of the ten clay discoidal whorls, nine are made of locally available red clay, and one of gray clay.

Seven of the whorls were classed as ellipsoidal (Fig. 45c). This style varies from subspherical to a thick discoidal shape. All the ellipsoidal whorls are made of clay; one is of gray clay, which does not occur locally. The only decorated spindle whorl belongs to this class. The design consists of two incised lines on the same side, which follow the inner and outer edges of the whorl.

Seven whorls fit into Haury's biconical class. In this group, the two cone shapes are symmetrical. All of the specimens are made of clay, one of extrinsic gray clay. Three are partially blackened but not fully fired (Fig. 45e, f).

TABLE 26

DISTRIBUTION OF SPINDLE WHORLS AND WORKED
AND DECORATED STONE

	Spindle Whorls	Painted Stones	Incised Stones	Beads	Shaft Smoothers	Worked Sandstone	TOTAL
NP	4	6	3				13
Ib			1				1
IIb		2	1		1		4
IIIa		2	1				3
IIId	1	1	1				3
IIIf			1			1	2
IIIg	1	1	4				6
IIIh		2	1			4	7
IIIi			1				1
IIIj			2				2
IVb			1		1		2
IVc	1		2	1			4
IVd	2						2
Va	1	2	2	1		2	8
Vb	7	2		1	1		11
Vc	1	1		1			3
Vd		2				2	4
Sur	5	2	1				8
TOTAL	23	23	22	4	3	9	84

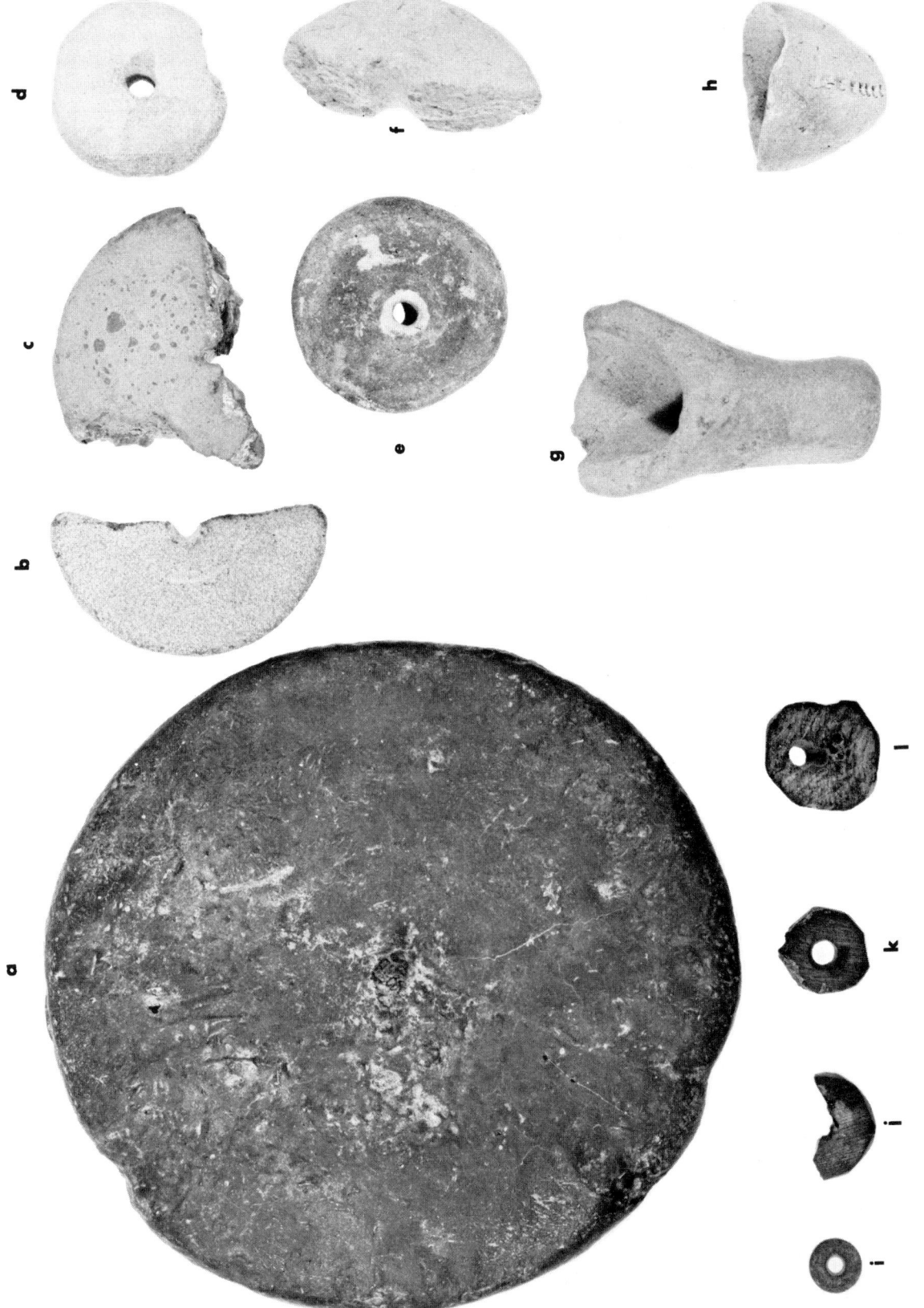

FIG. 45. Artifacts of clay, stone, and pitch: a, discoidal spindle whorl made of pitch (8.5 cm. in diameter); b, discoidal spindle whorl made of stone; c, ellipsoidal spindle whorl; d, decorated spindle whorl; e, biconical spindle whorl; f, fire-blackened spindle whorl; g, cornucopia-shaped object of clay; h, thimble-shaped object (decorated); i-k, beads made of stone; l, bead made of charcoal.

Worked and Decorated Stone

PAINTED STONE. Twenty-three examples of painted stone were recovered from Cowboy and Walters Caves. They were found from Stratum IIb to the surface. There was no specific, areal concentration of the stones; they ranged from the very front of the cave to the limit of the cultural remains excavated. The stones vary in length from 2.4 cm. to 19.0 cm., the average being 7.1 cm., and are from 0.3 to 3.4 cm. thick. Three pieces are oval or egg shaped with smoothed edges. Two are cigar shaped, but show no evidence of being worked except for the application of the painted designs. The remainder appear to be fragments of larger pieces.

The colors used in decorating the stones are red and black, or combinations of both, with one exception; this is a stone with white pigment markings. The most prominent design element is the stripe, which appears in some form on half of the stones--either horizontal, vertical, or angled, with different combinations of color. Five stones have black lines in no discernible pattern which could have been made with charcoal. A unique piece has two stripes combined with two rows of U-shaped designs in red pigment.

The three oval stones show a variety of patterns. The first, made of smooth, tan sandstone, has a small black stripe at the top, and the rest of the surface is covered with red ocher (Fig. 46a). Another has a black cross design at the top, and the bottom third is covered with black pigment (Fig. 46b). The last oval stone (Fig. 46c) has three red lines radiating from one end; the other half of the stone is covered with a pattern of lines and dashes.

The cigar-shaped stones are the only specimens with designs on both sides. The first, 10.5 cm. in length, has a number of short, black lines on the front arranged randomly except for three horizontal lines in the middle of the piece. The back, which is flat, has one line along its entire length, with alternating short lines nearly perpendicular to the long line (Fig. 46d, e). The second stone, 5.1 cm. in length (Fig. 46f, g), has a combination of red and black designs. At the top is a red cross with the four corners covered in black pigment which extends to the back. Five red stripes and then three black stripes encircle the rest of the stone.

Two large, thick pieces of stone were found which may have been used in preparing pigments. One has red ocher covering its surface and the other shows a yellow pigment under a coating of red ocher.

INCISED STONE. Twenty-two incised stones were recovered from the two caves. The decorations on the tabular stones were incised in two ways. The most common method was by scratching on the surface of a soft stone with a relatively harder tool. The second method, which has been described as "walking" a chisel-like tool along the surface (Schuster 1968:6), leaves a narrow zigzag pattern on the stone. This zigzag decoration is characteristic of incised stones from Utah, Nevada, and adjacent areas (Fig. 46h, i).

Thirteen of the Cowboy Cave specimens are scratched, and six are "chisel-walked;" all but one are sandstone, the exception being a shale piece recovered from Stratum IIIj. These artifacts range in length from 3.2 to 13.2 cm.; in width from 1.7 to 8.7 cm.; and in thickness from 0.1 to 1.0 cm. Nineteen of the 22 incised stones recovered are of known provenience, representing all excavated units at the site.

Archeological sites such as Danger Cave, Hogup Cave, and Swallow Shelter (Jennings 1957, Aikens 1970, and Dalley 1978) yielded numerous incised stones. The earliest Hogup Cave material came from strata dated 6190±110 B.P. Sixty-three specimens of incised volcanic tuff were recovered from Swallow Shelter; their proveniences, however, were not reported. One artifact from Stratum Ib at Cowboy Cave is the earliest recorded occurrence of incised stone from Utah and the adjacent areas, and there are 10 specimens from strata that date older than 6675±75 B.P. (Stratum IIIi).

Like other rock art, the incised

FIG. 46. Decorated stone: a–c, painted sandstone; d,e, same object (10.6 cm. in length) front and back; f,g, incised sandstone (front and back view); h,i, incised sandstone.

stone is difficult, if not impossible, to interpret. There is conjecture (Parsons n.d. and Newcomb 1976) that some rock art objects are associated with menstrual taboos or cult activities.

BEADS. Four beads were recovered (in addition to those reported in *Bone and Shell Material*). They vary in dia. from 0.7 cm. to 1.9 cm. Three of them are of black stone, possibly lignite, while one is made of charcoal, or wood which was later burned. The stone beads are round, with thin, flat sides and are well made. The charcoal bead is thicker than the other three and the hole is off-center (Fig. 45i, j, k, 1). All the beads were found in the upper levels of the caves.

SHAFT SMOOTHERS. Three examples of small, grooved stones, possibly shaft smoothers, were recovered. They vary in length from 3.0 cm. to 7.3 cm. and are made of local sandstone. Two stones have short, thin grooves near one edge, while the largest stone has a wide groove along its entire length. These stones were found in Strata IIb, IVb, and Vb.

WORKED SANDSTONE. Nine pieces of sandstone show evidence of being worked. Seven are of local red sandstone and two are of white sandstone. All the examples have smoothed sides and edges; three are made into disks, ranging in shape from oval to round. The largest disk is 8.9 cm. in dia. and 4.8 cm. in thickness. These pieces were found in Units III and V (Table 26).

Unfired Clay Objects

There were 144 specimens of unfired clay recovered from the cultural deposits in Cowboy and Walters Caves. Most of the artifacts appear to be waste fragments resulting from the manufacture of figurines. There are, however, 29 deliberately formed or decorated clay objects similar to those reported from archeological sites in the southwest (Kidder and Guernsey 1919, Morss 1954, Gunnerson 1959, Lipe 1960, Sharrock and Keane 1962, Wormington 1966, Amsden 1949, and others). They are categorized here descriptively as thimble-shaped,

loaf-shaped, cornucopia-shaped, or, in the case of 12 recognizable anthropomorphic forms, figurines. Proveniences appear in Table 27.

The thimble-shaped objects and cornucopia objects are similar to artifacts reported from Basketmaker sites (Amsden 1949), but the Cowboy Cave punctated loaf-shaped objects are not common. These particular specimens have rows of punctations decorating the pinched side of a loaf of clay.

CLAY FIGURINES. The few complete figurines (see Fig. 47) are of the "handle terminus" form attributed to the Fremont (Morss 1954:28). Some fragments show simple nose ridge modeling similar to Mogollon examples (Morss 1954:27) while other fragments are simply formed punctated figurines very similar to a specimen from Ventana Cave (Haury 1975: 360).

Clay figurines recovered during the Glen Canyon salvage program (Gunnerson 1959, Lipe 1960, Sharrock et al. 1961, and Sharrock and Keane 1962) were mostly of the pinched nose ridge form with punctate or appliqued eyes and mouth, and with either punctate or appliqued collars or necklaces. The Cowboy Cave figurines lack these detailed adornments. Ceramic chronologies and stylistic similarities were the bases for assigning BM II to BM III ages to southwestern clay artifacts from early excavated sites (Kidder and Guernsey 1919) as well as the proposed dates of 1150 B.P. to 750 B.P. to the Glen Canyon material. Thus, a Basketmaker/Fremont figurine trait concept has been perpetuated (Morss 1954: 27, 28).

More recent studies from the Lower Pecos area of Texas (Shafer 1975:151) and from northeastern California (Riddell 1960:58) report similar gynecomorphic cigar-shaped clay figurines dating from 4000 B.P. to about 3000 B.P. Although these objects are unlike the Cowboy Cave figurines, they lend credibility to early figurines in Utah.

Recent excavations at Sudden Shelter, which is near the Cowboy Cave site, produced clay artifacts of even greater antiquity than the figurines mentioned

TABLE 27

DISTRIBUTION OF UNFIRED CLAY OBJECTS

	Clay Figurines	Thimble-shaped Objects	Loaf-shaped Objects	Cornucopia Objects	Decorated Fragments	Undecorated Fragments	Random Clay Pieces	TOTAL
Ib							1	1
IIb	1		1					2
IIIb							1	1
IIIc							1	1
IIId	1						1	2
IIIi		1			3	1	9	14
IIIj	1							1
IIIk			1					1
IVa	3	1			3		10	17
IVb						1	2	3
IVc							1	1
IVd		1			1		3	5
Va							7	7
Vb	4			1		4	32	41
Vc	2						19	21
Vd					2	2	4	8
Sur			1		1		16	18
TOTAL	12	3	3	1	10	8	107	144

above. A gray figurine with a punctate chevron design, similar to some Cowboy Cave fragments, was bracketed between radiocarbon dates of 6670±180 B.P. and 6310±40 B.P. (Jennings et al. In press). The earliest date for figurines in the Cowboy and Walters Caves collection is 8875±125 B.P. This is derived from the association of the large figurine from Walters Cave (Fig. 47a) with a yucca fiber sandal which was radiocarbon dated. The Cowboy Cave provenience for a loaf-

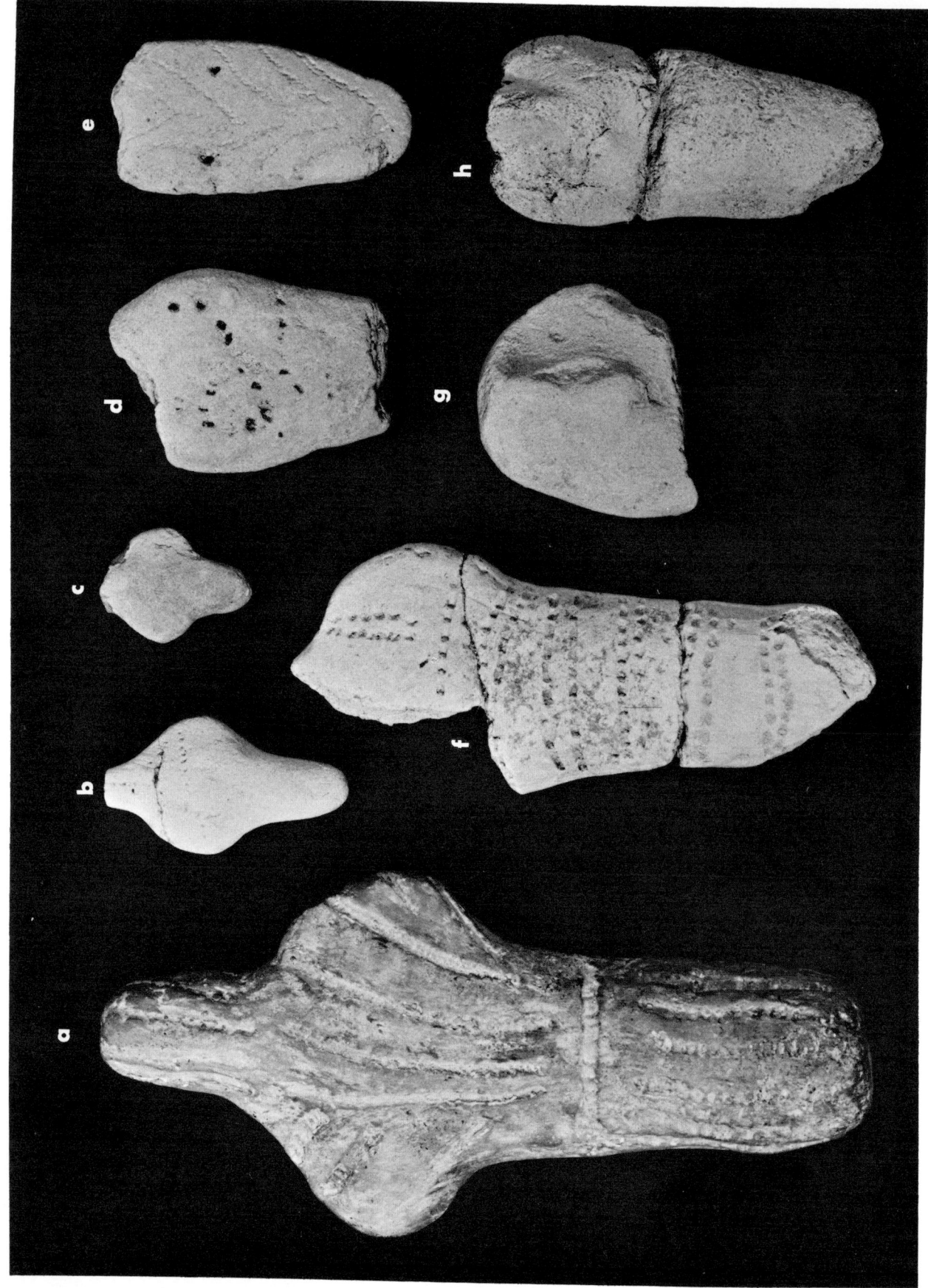

FIG. 47. Unfired clay figurines: a, whole (10.5 cm. in length); b,c, missing head; d, tatoo with vegetable fiber in it; e, with decoration on dorsal side also; f,g, head with nose ridge; h, whole figurine with nose ridge.

shaped object, a thimble **object,** three decorated fragments, and other clay fragments is Unit III, which dates between 6675±75 B.P. and 6390±70 B.P. These dates are almost identical to the radiocarbon dates that bracket the Sudden Shelter figurine (Jennings et al. In press). The evidence from Cowboy Cave and Sudden Shelter, therefore, challenges the continued designation of clay figurines as being only Basketmaker and Fremont clay artifact types.

PLANT MACROFOSSIL ANALYSIS

Peggy R. Barnett and Nancy J. Coulam

Plant macrofossils from Cowboy Cave were well preserved by the arid environment. Not only were carbonized plants recovered, but leaves, nuts and grasses were abundant in all occupational levels. Analysis was undertaken to determine the dietary importance of the plants to the aboriginal inhabitants.

METHODS

Of the 54 samples collected from both caves, 15 from Cowboy Cave were selected for analysis. Seven of the 15 samples were chosen to correspond to the strata from which coprolites were analyzed--IIb, IIIi, and IVc--in order to compare the results of these two studies (see *Appendix IX*).

Initially, four samples were processed through flotation using Streuver's (1968) method with modifications by the authors (Jennings et al. In press). However, it was determined that simply dry-screening the samples through a series of mesh sieves was a more efficient method of recovering the plant fragments and the remaining samples were processed in this manner. The first sieve, 4.699 mm. opening, captured large pieces of charcoal and the largest plant remains, i.e., leaves, twigs, bark, and nuts. The second sieve, 2.38 mm. opening, caught large seeds and smaller plant parts. The third and final sieve,

0.355 mm. opening, trapped the smallest seeds, plant parts, and debris. The final fine screen collection was sorted and identified using a 40X stereoscopic microscope. In addition to the plant macrofossils analyzed from the screenings and flotation, bulk samples of botanical material were cursorily examined for any additions to the plant list (Table 28).

Each plant taxa identified from the three screenings and flotation was recorded. Because the number and/or amount of each plant or plant fragment reflects nothing more than preservation for that particular sample, all tabulations are based simply on whether or not that taxa was represented in the sample, i.e., for any single sample each taxon may be identified a dozen or more times, but it was only counted as one occurrence. This insures that accidents of preservation are not mistaken for changes in cultural utilization.

RESULTS

The most frequently occurring macrofossil was grass, most of which could be identified only to the family. Other important components, in order of diminishing frequencies of occurrence, are *Juniperus* sp., *Chenopodium* sp., *Corispermum hyssipfolium*, and Cactaceae. The actual frequencies of these and all

TABLE 28

PLANTS REPRESENTED AT COWBOY CAVE

Family	Genus, Species	Common Name
AMARANTHACEAE	*Amaranthus* sp.	Amaranth
ANACARDIACEAE	*Rhus trilobata*	Squawbush
BERBERIDACEAE	*Berberis fremontii*	Fremont barberry
BORAGINACEAE	*Cryptantha* sp.	Cryptantha
CACTACEAE	*Echinocereus* sp.	Hedgehog cactus
	Opuntia sp.	Prickly pear cactus
CHENOPODIACEAE	*Chenopodium* sp.	Goosefoot
	Corispermum hyssipfolium	Bugseed
COMPOSITAE	*Ambrosia acanthicarpa*	Annual bursage
	Artemisia sp.	Sagebrush
	Dicoria brandegei	Dicoria
	Helianthus annuus	Common sunflower
	H. petiolaris	Narrowleaf sunflower
CRUCIFERAE	*Lepidium montanum*	Montana pepperweed
	L. sp.	Pepperweed
CUPRESSACEAE	*Juniperus osteosperma*	Utah juniper
	J. sp.	Juniper
CYPERACEAE	*Carex aquatilis*	Water sedge
	C. sp.	Sedge
EPHEDRACEAE	*Ephedra nevadensis*	Nevada Mormon tea
	E. torreyana	Torrey Mormon tea
FAGACEAE	*Quercus gambelii*	Gambel oak
GRAMINEAE	*Hilaria jamesii*	Galleta grass
	Oryzopsis hymenoides	Indian ricegrass
	Poa sp.	Bluegrass
	Sporobolus cryptandrus	Sand dropseed
	S. sp.	Dropseed
LEGUMINOSAE	*Astragalus mollissimus*	Wooly milkvetch
LILIACEAE	*Yucca* sp.	Soapweed
LOASACEAE	*Mentzelia* sp.	Blazing star
PINACEAE	*Picea pungens*	Blue spruce
	Pinus edulis	Pinyon pine
POLYGONACEAE	*Eriogonum* sp.	Wild buckwheat
ROSACEAE	*Amelanchier* sp.	Serviceberry
SALICACEAE	*Salix discolor*	Pussy willow
SANTALACEAE	*Comandra umbellata*	Common bastard toadflax

*No previous ethnohistoric use reported.

identified taxa are presented in Table 29. All of the listed plants were probably available within a short distance of the caves. The environment in the vicinity of Cowboy and Walters Caves provided a broad range of flora within each of three vegetation zones: the canyon bottoms in which the caves are located, nearby Spur Canyon and other tributary canyons, and the upland-flat grass communities above the canyons (see *Appendices VII, VIII*).

The examination of the bulk samples revealed little difference from the macrofossil results. Grasses were still the major component, followed by *Juniperus* sp., **Cactaceae**, and *Chenopodium* sp. The few additions were: a modern sunflower seed (*Helianthus annuus*), whole acorns (*Quercus gambelii*), whole cactus seeds (*Opuntia* sp.) and what appear to be corn inflourescenses (*Zea mays*). While some of these plants are available year round, most are available only in spring, summer and early fall (Table 30).

CONCLUSIONS

The flora identified from the macrofossil analysis (Table 29) differs from the modern floristic list (*Appendices VII, VIII*), indicating either a climatic change or, more probably, a vegetation change due to interference by man, i.e., the disappearance of some of the abundant grasses from the area near the caves by cattle grazing that began about 1870 (Cottam 1947). Some of the grasses recovered from the cave are today unavailable. Today *Muhlenbergia pungens* is a dominant grass in the grass communities on the flats back from the canyon rims. This species is rejected as food by range cattle and it might have replaced some more palatable range grass.

Another example of the disappearance of species from the environment of the caves is *Corispermum hyssipfolium*, which was recovered from every stratum—Ib through Vc—of Cowboy Cave. None of the early botanists in the West recorded it

and modern botanists (Kearney and Peebles 1964, Rydberg 1929) list *Corispermum hyssipfolium* as an introduced species. The macrofossil analysis and the coprolite analysis records, however, agree in confirming it as an **indigenous species**.

It is of interest to note that pinyon (*Pinus edulis*) first appears in Stratum IVc at about 3300 B.P., differing with the report in this paper by Spaulding (*Appendix II*), but agreeing with other evidence from the Great Basin on the appearance of pinyon (Madsen 1973). The late occurrence of this plant in the cultural macrofossil samples, however, may be due to accidents of preservation and does not rule out utilization by cave occupants at an earlier time.

The plant taxa that are named in ethnobotanic sources as food (Harrington 1967, Sweet 1962, Steward 1938, and Wyman and Harris 1951) are listed in Table 28. One cannot say, of course, that the aboriginal populations of Cowboy Cave used these plants in the same manner or at the same time. From the available edible plant inventory, however, the occupants apparently were temporarily camped in the caves in the spring, summer, and early fall months gathering wild plants.

TABLE 29

PLANTS IN DIMINISHING ORDER OF FREQUENCY
OF OCCURRENCE IN ALL SAMPLES

				Strata					
	IIa	IIb	IIIg	IIIi	IVa	IVc	Vb	Vc	Total
Number of Samples Analyzed	1	3	2	1	1	3	3	1	15
									Total Occur-rences
GRAMINEAE	1	3	2	1		3	3	1	14
Juniperus sp.	1	3	2	1	1	1	3	1	13
Chenopodium sp.		3	1			3	3	1	11
Corispermum hyssipfolium		2		1	1	2	3	1	10
Oryzopsis hymenoides	1	3		1		3	2		10
Sporobolus sp.		2		1	1		3	1	8
CACTACEAE		3			1	1	2		7
Sporobolus cryptandrus	1		2			2			5
Pinus edulis						2	2	1	5
Helianthus petiolaris		2				1	1		4
Cryptantha sp.			1	1		1			3
Opuntia sp.		1	2						3
CRUCIFERAE		1	1				1		3
Juniperus osteosperma	1	1				1			3
EPHEDRACEAE						2	1		3
Berberis fremontii		1				1			2
Artemisia sp.			1				1		2
Dicoria brandegei		2							2
Carex sp.						2			2

TABLE 29 (continued)

PLANTS IN DIMINISHING ORDER OF FREQUENCY

OF OCCURRENCE IN ALL SAMPLES

	Strata								
	IIa	IIb	IIIg	IIIi	IVa	IVc	Vb	Vc	Total
Ephedra nevadensis		1			1				2
Quercus gambelii	1		1						2
Hilaria jamesii						1	1		2
Amaranthus sp.							1		1
Rhus trilobata			1						1
BORAGINACEAE							1		1
Echinocereus sp.							1		1
COMPOSITAE							1		1
Ambrosia acanthicarpa			1						1
Helianthus annuus						1			1
Lepidium montanum							1		1
Carex aquatilis			1						1
Ephedra torreyana		1							1
Poa sp.						1			1
Astragalus mollissimus			1						1
Yucca sp.							1		1
Mentzelia sp.							1		1
Picea pungens							1		1
Eriogonum sp.		1							1
Amelanchier sp.		1							1
Salix discolor						1			1
Comandra umbellata			1						1

TABLE 30

SEASONAL AVAILABILITY OF PLANTS AS FOOD

	Jan	Feb	Mar	Apr	May	Jun	Jul	Aug	Sep	Oct	Nov	Dec
GRAMINEAE												
Juniperus sp.				├	─	─	─	─	─	─	─	┤
Chenopodium sp.				├	─	─	─	─	─	┤		
Corispermum hyssipfolium							├	─	┤			
Oryzopsis hymenoides					├	─	─	┤				
Sporobolus sp.						├	─	─	┤			
CACTACEAE	├	─	─	─	─	─	─	─	─	─	─	┤
Sporobolus cryptandrus						├	─	─	─	┤		
Pinus edulis	├	─	─	─	─	─	─	─	─	─	─	┤
Helianthus petiolaris						├	─	─	┤			
Cryptantha sp.				├	─	─	─	┤				
Opuntia sp.	├	─	─	─	─	─	─	─	─	─	─	┤
CRUCIFERAE												
Juniperus osteosperma	├	─	─	─	─	─	─	─	─	─	─	┤
EPHEDRACEAE												
Berberis fremontii				├	─	─	─	┤				
Artemisia sp.				├	─	─	─	─	─	─	┤	
Dicoria brandegei						├	─	─	┤			
Carex sp.						├	─	┤				
Ephedra nevadensis	├	─	─	─	─	─	─	─	─	─	─	┤

TABLE 30 (continued)

SEASONAL AVAILABILITY OF PLANTS AS FOOD

	Jan	Feb	Mar	Apr	May	Jun	Jul	Aug	Sep	Oct	Nov	Dec
Quercus gambelii						X	X	X	X	X		
Hilaria jamesii					X	X	X	X	X			
Amaranthus sp.				X	X	X	X	X	X			
Rhus trilobata			X	X	X	X	X	X	X			
BORAGINACEAE												
Echinocereus sp.	X	X	X	X	X	X	X	X	X	X	X	X
COMPOSITAE												
Ambrosia acanthicarpa							X	X	X	X		
Helianthus annuus			X	X	X	X	X	X	X	X		
Lepidium montanum					X	X	X	X	X			
Carex aquatilis						X	X	X				
Ephedra torreyana	X	X	X	X	X	X	X	X	X	X	X	X
Poa sp.						X	X	X				
Astragalus mollissimus					X	X	X	X				
Yucca sp.					X	X	X	X				
Mentzelia sp.						X	X	X	X			
Picea pungens	X	X	X	X	X	X	X	X	X	X	X	X
Eriogonum sp.						X	X	X	X			
Amelanchier sp.				X	X	X	X	X				
Salix discolor			X	X								
Comandra umbellata						X	X	X				

THE OCCURRENCE OF PINYON PINE AT COWBOY CAVE

Nancy J. Hewitt

In the past decade or more, there has been discussion concerning the recency of aboriginal usage of pinyon pine in the Great Basin. Evidence of pinyon pine from Cowboy Cave includes not only nuts, needles, and chunks of pitch, but artifacts displaying aboriginal utilization of this valuable resource (Table 31). Among them are pitched basketry fragments, pitch spindle whorls, Gypsum projectile points with pitch adhering where they were attached to a shaft, and several sticks with pitch on one end which may have been used to caulk baskets, making them virtually watertight. Except for two anomalies, none of these artifacts or macrofossils occurs prior to Unit IV.

Radiocarbon dates indicate that cultural Unit IV spans at least from 3635 to 3330 B.P. (Tables 2, 3), and the association of Gypsum points suggests that the initial occupation of Unit IV could not have been earlier than 4600 B.P. (see *Chipped Stone Projectile Points*). The sudden appearance of macrofossils and artifacts utilizing pinyon pine in Unit IV deserves comment, because it suggests that either pinyon was not locally available during the Unit II and Unit III occupancies, or that it was not discovered and exploited at Cowboy Cave prior to about 3500 B.P.

TABLE 31

DISTRIBUTION OF PITCH, ARTIFACTS USING PITCH, AND
BOTANICAL REMAINS OF PINYON PINE

	NP	Unit: III		IV				V				Sur	TOTAL
Stratum:		i	j	a	b	c	d	a	b	c	d		
Chunks of Pitch	2	1		1		1		1	3	2		2	13
Pitched Basketry	1						1	1					3
Sticks with Pitch								1	3	1			5
Gypsum Points				4	1	7	3	6	5				26
Botanical Remains of Pinyon		1				2			2	1			6
Pitch Spindle Whorls							1		1				2
TOTAL	3	2		5	1	10	5	9	14	4		2	55

ANIMAL SKIN BAGS

Frank W. Hull

INTRODUCTION

Originally this report was to be concerned only with the study of the animal skin bags found during the testing and excavation of Cowboy Cave. Review of the records on file at the University of Utah Archeology Laboratory and specimens from the collections of the Utah Museum of Natural History, however, lead to an extension of the discussion to include the bags and pouches from other University of Utah expeditions in Utah and adjacent areas. Ten bags in all were brought together for description and analysis. Various other skin bags are also mentioned in order to provide a broader comparative basis for discussion.

In the absence of a model for classification of these artifacts, descriptive categories were established, based on the nature of the contents of the bags. They are: personal item containers, tool kits, paint kits, bulk storage bags, and medicine bundles and bags. In order to provide some consistency in the usage of animal and plant names, the terminology of Durrant (1952) and Beetle (1970) is used throughout, except where consultants are noted.

Many of the materials examined required specialized knowledge for identification. I acknowledge with appreciation Earl Crum, a Shoshoni of Owyhee, Nevada, who identified the Cowboy Cave animal skin bag as a fawn's head; Lois Arnow of the Garrett Herbarium, University of Utah, who identified seeds of the Cowboy Cave necklace; William Behle, University of Utah Biology Department, who provided the scientific names of many of the feathers of the medicine bundle; and Baldomero Olivera, who examined the exotic shells and shark's tooth artifacts attached to the medicine bundle.

Personal Item Containers

Items for personal decoration and other valued personal objects were carried or stored in these containers.

COWBOY CAVE ANIMAL SKIN BAG. A small, soft hide bag (ca. 18 cm. long) was recovered from Cowboy Cave in Wayne County, Utah, during testing of the site in October, 1973 (Fig. 48). It came from an upper stratum under a fossil wood rat (*Neotoma* sp.) nest, which casts some doubt on its provenience. A wad of grass (*Sporobolus*), evidently used as padding in the bag, yielded a radiocarbon date of 3330±80 B.P. Because the date was compatible with other dates from Unit IV, the bag was assigned to that unit.

The leather was dry and extremely fragile, and it was necessary to soak it in distilled water until it was pliable enough to open. At this point, several features of the bag could be seen for the first time. It had been fashioned from an unsplit animal skin, later identified as a fawn's head. Leather patches were sewn over the natural

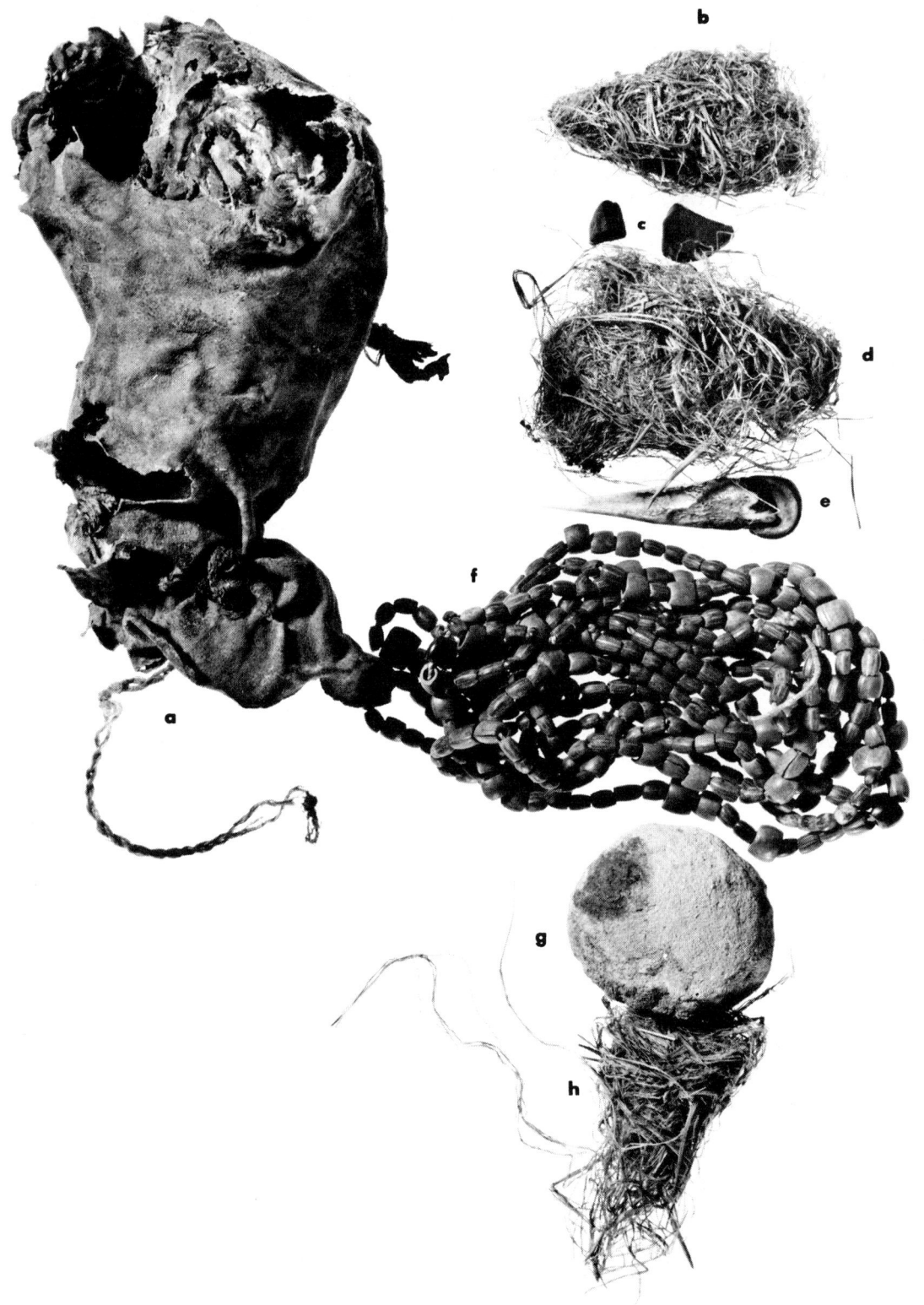

FIG. 48. Fawn's head pouch and contents (in order of removal): a, pouch; b, fiber pad; c, hematite stones; d, fiber pad; e, bone awl (7.5 cm. in length); f, seed necklace; g, stone ball; h, fiber pad.

openings of the animal skin with fine yucca twine (*Yucca angustissima*) and also over a hole worn near the bottom of the bag. The top of the bag had a twisted fiber string tied through four holes attaching the cord permanently to one side of the bag. The string was then wrapped twice around the top of the bag, and the loose end tucked under the wraps. In the creases of the leather where the fiber string had puckered the bag, the hair was still intact; most of the hair was worn off the main part of the pouch.

The contents were removed and recorded. On top was a dropseed grass mat (*Sporobolus cryptandrus*) over an irregular piece of sandstone that measured 5.8 by 5.3 by 3.6 cm. The stone exhibited no wear patterns or man-made alterations.

Next to be removed from the pouch was a necklace made from three different plant seeds that had been strung onto twisted vegetable fiber cords to make four complete strands and a fifth half-strung with beads. The large beads were seeds of *Prunus americana* (wild plum) that is today native to northern Utah, Wyoming, Colorado, and Idaho. Smaller, striped beads--elliptical seeds that had the ends ground off--were made from seeds of *Elaeagnus argentea*, commonly called silverberry. They are of the same genus as Russian olive, and, like wild plum, are indigenous to northern Utah and the tier of states to the north. The smallest beads were seeds of *Juniperus cf. osteosperma*, native to Utah. The strands ranged from 70 to 75 cm. long.

Underneath the bead necklace a well-made bone awl lay on another pad of vegetable fiber. The awl, 7.5 cm. long, was made from a split *Antilocapra americana* (antelope) metatarsal. The unaltered "palm" end of the bone made the identification possible.

Two pieces of red ocher were cushioned between a third wad of vegetal fiber on the bottom of the leather bag and the middle pad. The larger piece, about 1.5 cm. across, was tabular in shape. The inside of the pouch was stained with red ocher, indicating repetitive storing of this substance.

The lack of what might be termed exotic items tends to rule out a "medicine bundle" classification for the Cowboy Cave bag. Though one could possibly ascribe ceremonial significance to the use of the nearly unaltered skin from a fawn's head, it more likely represents a practical use of a highly valued resource, as was suggested by the patched and repatched leather artifacts from Danger Cave (Jennings 1957:220).

Pigment and staining were important elements of the pouch. Historically, red pigment was used on articles of clothing as well as for body adornment (Wheat 1967). These and the other contents--a small piece of sandstone, a seed necklace, a bone awl, and three fiber pads--were most likely personal items of some value and utility to their owner.

SAND DUNE CAVE BAG. Sand Dune Cave, located on the northeast side of Navajo Mountain, was excavated in 1961 (Lindsay 1968:30) as part of the Glen Canyon Project. A bag formed of a whole dog skin was recovered from the cultural deposits. The following list of its contents, in order of removal, is from Lindsay (1968:42):

A. Twined bag, partially folded upon itself and tied with cord. It, in turn, contained:
  1. A small hide bag containing 13 gaming pieces.
  2. Two turkey feather *pahos*.
  3. Two red-winged blackbird wing skins tied together.
  4. One bone awl.
  5. One bone tube.
  6. One flat river cobble used as a flaking stone.
B. Small bag of prairie dog skin, folded over at the mouth and tied with cordage. It contained:
  1. Eighteen dart point blanks.
  2. Two lumps of uranium ore, fused together.
  3. A wooden flaker.
C. Small bag of prairie dog skin, gathered at the end and tied with split yucca leaf. It contained:

1. Eight cylindrical gaming sticks of horn.
2. One conical object of horn.
3. Two small spheres of sulphur.

D. Bundle of six dart foreshafts with stone points, a red-tailed hawk feather, and a piece of sinew, tied with cordage.

A cultural affiliation of Basketmaker II was assigned due to stratigraphic association with Desha Complex-Basketmaker II open-twined sandals. The dog skin bag is slightly more recent than the radiocarbon date of the sandals (7000 to 8000 B.P.).

All together eight animal skin bags were recovered from Sand Dune Cave, all of which were from Basketmaker II levels. The 1916–17 explorations into northeastern Arizona (Guernsey and Kidder 1921) recovered similar Basketmaker caches in a badger (*Taxidea taxus*) skin bag.

Tool Kits

Bone, stone, or wooden material used for the manufacture of tools or as implements in daily subsistence activities and stored with similar objects, constitutes a tool kit.

WINNEMUCCA LAKE ANIMAL SKIN BAGS. The records in the Archeology Laboratory at the University of Utah contain photographs and some description of skin bags discovered by an amateur near Dry Winnemucca Lake, Nevada, which fit this category. Lacking access to the artifacts, this discussion is based on the available records, which provide some useful comparative data.

The Winnemucca Lake find is comprised of two skin bags filled with artifacts. The first bag was apparently the body portion of a small animal with the area between the hind legs sewn across to form the bottom of the bag. None of the small openings left by the removal of the legs were sewn or tied; a flap of skin which extended past the region of the ears was used for closure. A photograph of the contents shows a hafted knife; a length of cordage intertwined with feathers; a short piece of worked wood which resembled a fragment

of a bow or atlatl charred on one end; and 20 or more lanceolate and notched projectile points or blades. The artifact collection from the second, smaller bag shown included seven dart shafts with attached points and a fairly long stone blade (ca. 15 cm.) wrapped more than half its length with rawhide or leather. Also pictured is a long ivory-colored, awl-shaped object resembling a shuttle, with ears or spurs on the wider end and a small hole drilled in the other. There is a natural chevron pattern visible, which is consistent with its tentative identification (in an accompanying letter) as "of mastodon material, perhaps tusk."

Little can be said about age and cultural affiliation except from the presence of diagnostic point types. Eastgate expanding stem, Elko Corner-notched, and possible Humboldt projectile points, as well as a Pinto Sloping-shoulder dart point, can be recognized without difficulty. Aikens (1970:55) placed the Eastgate expanding stem and Elko Corner-notched points in a rather long continuum occurring from about 4450 B.P. to 600 B.P., and 7950 B.P. to ca. 600 B.P., respectively. Holmer (*Chipped Stone Projectile Points*) has more precise date ranges for the Elko series. The Pinto series at Hogup Cave was found to occur in the fill from 8350 B.P., terminating about 3300 to 2600 B.P. The Humboldt points at Hogup persisted in the cultural deposits from 7250 B.P. to 2600 B.P.

Paint Kits

BENCHMARK CAVE PAINT KIT. The University of Utah excavation of Benchmark Cave (Lipe 1960) produced a paint kit that has been fully described by Lipe (1960:186, Fig. 57). The small pouch, made from the body skin of a rock squirrel (*Citellus variegatus*) contained a smaller, tanned inner pouch that held the pigment and applicator. Cordage used in the tie string, thread, and on the applicator, is cotton of S-twist construction.

Due to the relatively late introduction of cotton (Howard 1968), this paint kit probably represents the occupancy of Benchmark Cave attributed by Lipe as early Pueblo III (950 to 850 B.P.). Guernsey and Kidder (1912) report a similar bag containing a "brilliant green powdered pigment" from Basketmaker cultural deposits in northeastern Arizona.

Bulk Storage Bags

BENCHMARK CAVE POUCH. Several well-preserved skin artifacts were recovered in the 1958 Glen Canyon excavations (Lipe 1960:188). Among them was a deerskin pouch filled nearly to the bursting point with cottonseeds (*Gossypium* sp.). The tubular, slightly tapered pouch (Lipe 1960, Fig. 54) was 23 cm. long and about 16 cm. in circumference. It was fashioned from a single piece of deerskin that was sewn on the bottom and up one side with sinew in an overcast stitch and turned inside out. A small wad of fiber had been used to plug the narrower open end.

Stratum III at Benchmark Cave, which contained the cottonseed pouch, also held Kayenta affiliated ceramics (Tsegi Orange ware) which date 950 to 850 B.P. (Lipe 1960:99).

REHAB CENTER HIDE BAG. Rehab Center was one of the several excavations carried out by the University of Utah during 1961 excavations in the Glen Canyon area (Sharrock et al. 1963). The report included a description of a bag approximately 35 cm. by 50 cm. in circumference that was made from the hide of a young fawn, hair side out. A kilo of "light gray powdery material primarily composed of quartz and gypsum, plus small amounts of mica, calcite, and ground potsherds" --perhaps temper for pottery--was inside the bag.

Pottery types of Tusayan Black-on-White, Tsegi Orange and Tusayan Polychrome, attributed to Pueblo III occupancy (ca. 800 to 650 B.P.), were found in close association with the hide bag, although not in a tightly controlled stratigraphic sequence.

COWBOY CAVE CORN BAGS. Two animal skin bags filled with an early variety of corn (*Appendix IV*) were recovered

from Cowboy Cave (Fig. 16). The corn was radiocarbon dated at about 1600 B.P. The smaller of the two bags was fashioned from the body of *Marmota flaviventer* (marmot) with the neck part of the animal's skin used for the opening of the bag. The front legs of the animal were inverted, sealing off those openings; the posterior had been split from leg to leg and the hide sewn up using a running stitch. Two-ply S-twist yucca fiber thread was used here, and also in placing a small patch on the ventral side near the bottom of the dressed bag. A human hair tie string of four-ply S-twist construction secured the open end. The bag, which measured 27 cm. in length by 40 cm. in circumference, was filled with 1.485 kg. of corn.

The larger bag, also of marmot, was 35 cm. in length and 59 cm. in circumference. It contained 4.342 kg. of corn of the same variety found in the smaller bag. This bag had deteriorated and had been patched at the bottom with a piece of leather. The legs had been tied with Z-twist yucca fiber twine to seal the openings. Similar cordage was used to sew the patches in place using running stitches.

The dressed bags were constructed from the whole body portion of the animals that, judging from their condition and softness, had been altered to some degree by tanning after most of the hair had been removed. The hair side was out, displaying traces of russet to cinnamon-buff marmot fur around the edges and on the leg portions.

Medicine Bundles and Bags

The nonutilitarian nature of the contents--especially the presence of exotic items such as crystals, feathers, fossils, teeth, turquoise, and other objects that are found in historically documented ritualistic paraphernalia--distinguishes this class of artifacts.

HOSTEEN CANYON BUNDLE. Hosteen Canyon, Tsegi-ot-Sosi Canyon, northern Arizona, was the focus of archeological investigations in 1909 by the University of Utah (Turner 1962). Several well-preserved artifacts were added to the collections from that expedition, one of which was a bundle encased in an animal skin that was reexamined for this study (Fig. 49).

The Hosteen Canyon bundle had the shape of an ear of corn with the stalk end contoured by wrapping the animal skin very tightly with Z-twisted cotton fiber cords. The body portion of a prairie dog (*Cynomys cf. gunnisoni*), hair side out, formed the container for numerous objects. Overall length of the unopened bundle was 32 cm., and it measured 21 cm. in circumference. The husklike outer covering had been trimmed off squarely; a buckskin thong encircled the bundle two and a half turns near the open end.

Removing the wraps revealed another pouchlike bundle only slightly smaller (23 cm.) than the outer bag. This one—a flat piece of buckskin gathered to form a pouch—enclosed a bundle of feathers joined by a very delicately made yucca fiber cord. A piece of deer skin, hair intact, and a slender quartz crystal 2 cm. in length were also attached. All of this was carefully wrapped and held together with a buckskin thong.

Various elements of the feather bundle were identified as follows:
1. The single feather on the end of the cord was a **tail** feather of a ladder-backed woodpecker, *Dendrocopos scalaris*.
2. The bright red patch was from the crown of a ladder-backed woodpecker, *Dendrocopos scalaris*.
3. A single detached feather and several tufts of long, soft, barred feathers were from a great horned owl, *Bubo virginianus*.
4. The bright, deep blue feathers, long and short, were from a western bluebird, *Sialia occidentalis*.
5. The bright yellow feather was from a western tanager, *Piranga ludoviciana*.
6. The small, blue gray feathers were from a scrub jay, *Aphelocoma coermescens*.
7. A small, pale gray feather was from a pinyon jay, *Gymnorhinus cyanocephalus*.
8. The long, soft, brown feathers (some chopped off short) were from a golden eagle, *Aguila chrysaetos*.
9. The combination red orange feather was probably from a red crossbill, *Loxia curirrostra*, but could be parrot.

Below the feather bundle was an assortment of articles loose in the bottom of the outer skin pouch. Two more small leather bags, a small pouch made from a maize leaf (*Zea*), a tabular piece of hematite 1.5 cm. by 0.5 cm., three small layered or banded stones, kernels of corn, squash seeds, a crystal with a central groove as if for attachment, *Amaranthus* sp. seeds, a small stone that might be a portion of a trilobite fossil, and a fossil shark tooth were recorded. The medium-sized bag was fashioned from a piece of a rodent skin tied with S-twisted yucca fiber cord similar to the large outer skin pouch. The small, shiny, black *Amaranthus* seeds were found to have spilled out of a tear in the fourth bag, a small, gathered scrap of very soft buckskin.

Photographs were taken as the feather bundle was carefully unwrapped. A rather loose loop of yucca cordage threaded through three squash buttons was teased off without untying, as was the loop with the quartz crystal. Four small feather bundles held together by four and a half wraps of buckskin thong formed the larger one.

One minor bundle of feathers contained a dense object previously detected by X-rays. Probing with tweezers and dental tools located a rigid central shaft with a fossil shark's tooth attached securely with a sinew. This bundle was thus a miniature version of the major feather bundle.

A second, smaller bundle had two small shells strung onto the fastening cordage; X-rays indicated the presence of three or four other shells in the bundle. Alteration of the visible shells by nature or man had worn away many of

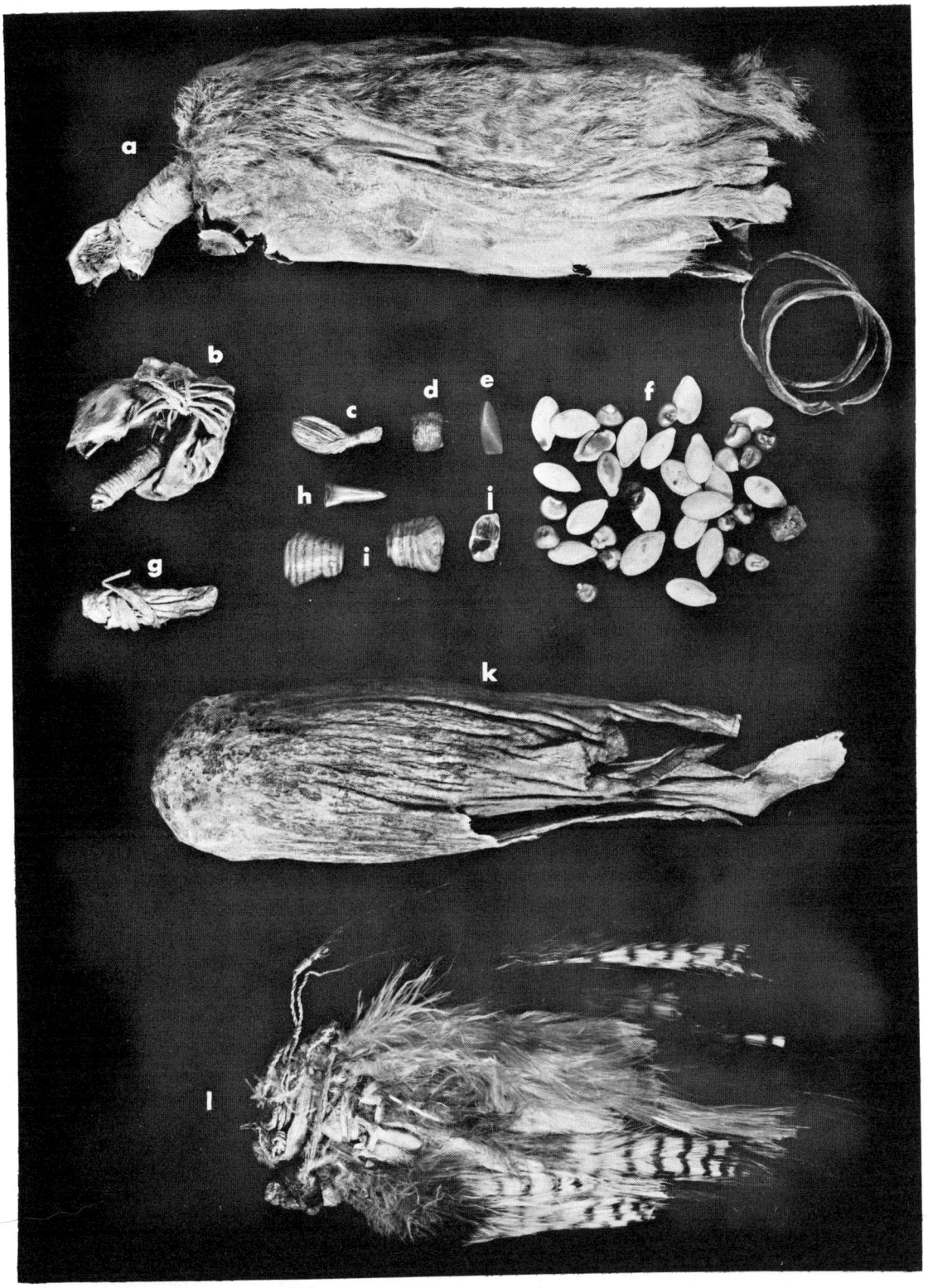

FIG. 49. Contents of Hosteen Canyon bundle: a, outer animal skin pouch (approx. 25 cm. in length); b, leather pouch filled with small seeds; c, plant leaf pouch; d, trilobite fossil; e, hematite stone; f, squash and corn seeds; g, leather pouch; h, shark tooth; i, banded stones; j, quartz crystal; k, leather pouch that enclosed feather bundle; and l, feather bundle.

the distinguishing characteristics, however, they are probably from the family
Olividae or *Mitridae*, perhaps *Olivella*
sp.

A third minor feather bundle had no
bone or shell objects attached. Like
the others, however, it was made of a
variety of feathers individually attached to a cord by a series of half
hitches placed around the proximal end
of each feather.

The last of the four minor bundles,
assembled like the others, was composed
entirely of eagle feathers.

The yucca fiber cords on the Hosteen
Canyon bundle were constructed with an
almost equal number of Z- and S-twists;
all of the cotton twine was Z-twist.

Because of its exotic and nonutilitarian contents--feathers, crystals, a
shark's tooth--the Hosteen Canyon bag
was classified as a medicine bundle.
The careful shaping of the pouch to
resemble an ear of corn, and the small
number of *Zea* (corn) and *Cucurbita* sp.
(squash) seeds argue in favor of ritual
rather than economic or subsistence
usage. The presence of cotton twine
indicates the bundle is younger than
1250 B.P. (Howard 1968:13).

SUMMARY

Hide bags and pouches like those
reported here are rarely found in excavations (Aikens 1970:112; Dalley 1978;
Jennings 1957:220; Wheat 1967:6). Their
preservation is possible only when these
normally perishable artifacts are sealed
in cultural deposits of very dry caves
and shelters. The recovery at Cowboy
Cave of three such bags gave impetus to
this broader analysis of skin bags from
Utah and adjacent areas.

Of the ten bags available for study,
nine had clear utilitarian functions:
four of them were storage bags; two contained hunter's tools; two held collections of personal items (though, to be
sure, the use of the individual objects
is not always clear); and one held the
pigment and paraphernalia of an aboriginal painter. Historic accounts (Wheat

1967) confirm the use of pouches for
carrying paints and other treasured
items.

The tenth bag, on the other hand,
(the Hosteen Canyon feather bundle) contained no objects of routine economic
importance. Instead, the quartz crystal,
shark's tooth, exotic feathers, and
shells are comparable to the contents of
a medicine bag described by Howard (1968:
13) that was found near Mesa Verde. Bond
(1941:26) describes the use of feather
bundles by modern Navajo medicine men as
part of a healing ritual. The feather
bundles from Hosteen Canyon are probably
from such an ancient medicine man's collection of healing devices.

SUMMARY

The foregoing pages constitute a descriptive report with relevant comparative and interpretive comments by a number of authors on the contents of Cowboy Cave. On the face of it, Cowboy Cave is merely another Western Archaic site with its affiliations largely toward the Great Basin, particularly in the earlier levels. It contained many interesting, even unusual artifacts; it has a challenging internal history and was pleasurable to excavate. However, as analysis proceeded, it was clear that there was more to be learned than originally met the eye.

Most sites can be interpreted on two quite different levels. Any individual site is always interesting and useful intrinsically. That is to say, it is unique, as all sites are unique, and provides its own history. Usually, too, there are unexpected and useful scraps of information nowhere else encountered in sites of the same or related cultures. But, in final analysis, the long-range importance of any site derives from whatever contribution it makes to an enriching of the understanding of an areal or regional cultural manifestation. In that aspect, Cowboy Cave appears to have been both timely and useful.

On the level of the intrinsic, site-specific, pieces of new information, several contributions can be cited. Among the artifacts, one can mention the twig figurines. Cowboy is among the few caves where the so-called Grand Canyon twig figurines occur in conjunction with the more mundane artifacts of daily life. Some twenty of the little animals occurred in the debris of living; of great interest is the fact that they fall in the 3500 B.P. time range, the same as the better known ones from Grand Canyon and westward. Interestingly, they are actually somewhat more complex in construction, in that they involve a final finishing technique nowhere else noticed. As Schroedl (1977) has suggested, they would appear to be more secular than sacred, a role assigned to earlier finds because they frequently occur in caches with no direct association with other artifact material.

Equally interesting and unexpected are the many painted and incised stones; these are somewhat rare in the archeological record, but occurred here in abundance. The evidence is that the practice of painting pieces of tabular sandstone characterized the Cowboy Cave cultures from the beginning. The incised stones are less rare, since they have been found at numerous eastern Great Basin sites. Despite Schuster's (1968) conviction that these incised stones are associated with a cult, and that each, if properly interpreted, represents an anthropomorph, the Cowboy Cave incised stones reveal no pattern subject to such an interpretation. They are, however, generally similar to the ones from Promontory, Hogup, and Danger Caves.

In the flint study, it becomes apparent once more that the so-called "dart points" were beyond doubt used as knives as well. Weder's study of edgewear and striations emphasizes this dual function; some points must have been made for use as knives and hafted on short, sturdy handles. Some, on the other hand, appear to have been used as projectile points.

The clay objects are also unusual. Among these crude fabrications of unbaked clay are cup-shaped or conical objects resembling thimbles; some solid pellets of clay pinched to a ridge on one side, called punctated loaves; a cornucopia-shaped object (Fig. 45g), and, predominantly, anthropomorphic figures (Fig. 47). They occur from the very earliest strata all the way to the final occupancy after the time of Christ. The little figurines are nowhere duplicated in the archeological record. Nor are they readily comparable to the Basketmaker figurines, or the later Fremont figurines, although they resemble, to some extent, the "handle terminus" forms described by Morss (1954). Their occurrence over such a long time range is new distributional data which may eventually prove to be of some importance.

The stability, or at least continuity, of artifacts displayed by the twig figurines, the clay figurines, and the painted and incised stones requires further study and explanation. It is particularly interesting in view of the long breaks in the occupational record: three major hiatuses occurred, each in excess of a thousand years.

Aside from the artifacts, the excavations at Cowboy Cave yielded other interesting and unusual kinds of information. As one example, one can mention the large quantity of corn in the cache. The kernels there are all plump and well formed; one can only conclude that the several kilos were carefully bagged and cached for seed corn. Evidently corn was being cultivated during Unit V, the final period of occupancy. The corn

itself is interesting in being a different variety from the corn usually called Basketmaker. It falls within the same time range as the corn from Clydes Cavern, i.e., about 1500 years B.P. In general, the occurrence of corn at Cowboy is compatible with the timing of its appearance in the Four Corners area. However, because both Cowboy Cave and Clydes Cavern fall much further north than corn has previously been reported, the Cowboy Cave cache holds somewhat more importance than it might have otherwise. (It should be noted that the radiocarbon ages of the corn kernels refute an earlier date initially ascribed to the corn in Jennings, 1975).

Another interesting botanical circumstance is the persistence of *Corispermum hyssipfolium*, a grass generally believed to be an introduced exotic because early botanists of the West did not note it in their original studies. The discovery that *Corispermum hyssipfolium* macrofossils characterized the Cowboy Cave deposits from the very beginning completely refutes the belief that it is an introduced species.

Equally interesting botanically is the lack of macrofossils, seeds, or gum of the pinyon until after 4000 B.P. There has been some discussion of the recency of pinyon in the archeological record over the Great Basin. Some authors (Madsen 1973 and Bettinger 1977) explain the lack of evidence of pinyon in the early western Archaic manifestation quite differently. Madsen avers that *Pinus edulis* was absent from southern Nevada until after 4500 B.P. Bettinger, on the other hand, would have it that the cultural utilization of the pinyon as a staple food represents a recent adaptation due to population increase by the Shoshoni-speaking tribes of the Great Basin. There is no doubt, however, that in the Rocky Mountains, pinyon appears throughout the prehistoric record, including macrofossils in the dung layer at Cowboy Cave, approximately 12,000 B.P. This scrap of information is difficult to assess at the moment.

Another interesting piece of botanical evidence is dietary. During Units

II through IV, the most frequent macro-fossils were grass. This fact, along with the paucity of animal bones, led early to the conclusion that the occupancy of the cave was confined to the spring, summer, and early fall months, and that the entire site was a special use area or location where the harvesting of wild vegetable foods was the primary activity.

An interesting change, however, occurs in Unit V. Here the chenopod/amaranths were the dominant economic grain. Whether to interpret this as a change in cultural preference, or the introduction of the amaranth as a companion plant with corn is not clear. It appears to be well known (Yarnell 1965) that the amaranth was introduced along with corn, and is originally a species from much further south. The importance of the amaranth as an aboriginal grain is generally accepted. Its use today as a ceremonial food is documented among the Aztecs and the Hopi, for example. Evidently, *Corispermum* or bugseed, mentioned earlier, was an equally important food grain up through Unit IV.

Some discrepancies occurred in the identification of the botanical samples from the leaf layer in Stratum IIa. Albee of the Garrett Herbarium, University of Utah, and Spaulding, Geochronology Laboratory, University of Arizona, differed on the identification of a few species in the collections. No effort was made to resolve the discrepancies; both series are listed, however, in *Appendix VI*.

Moving now from the unique and the site-specific items of interest to the implications of Cowboy Cave data in the wider regional sense, one can mention the close and clear artifactual similarities between the so-called Desha Complex of northern Arizona near Navajo Mountain. On several counts, including the characteristic sandal, the basketry manufacturing techniques, and certain other classes of artifacts, the Cowboy Cave material extends and lends credence to the Desha material. It is also likely that the Cowboy Cave evidence helps with the understanding of the

Desha Complex because at the two type sites (Lindsay et al. 1968), there was some doubt in the minds of the excavators as to whether the upper layers of both Sand Dune and Wind Devil (Dust Devil) Caves may have been disturbed. The impression one gets from Cowboy Cave is that the upper layers, not only at Cowboy, but at the Desha caves as well, were not necessarily Basketmaker II, as believed by Ambler, but were merely terminal Archaic which, in the Four Corners area, was labelled Basketmaker II. It seems more economical to interpret the upper layers of Desha and Unit V of Cowboy Cave as literally a terminal Archaic transition or base out of which the classic Basketmaker II as defined finally developed. A number of persons consulted about the Unit V material indicate that it has a "Basketmaker feel" but is not a pure manifestation of what has come to be known as Basketmaker II in the Four Corners area. In any case, it must be reiterated that the Basketmaker I substratum, so often referred to as "hypothetical," is not hypothetical at all. It is simply the late Archaic manifestation observed in the Southwest, in the Great Basin, and more recently, in the Colorado Plateau.

One of the most important Cowboy Cave findings is that the chipped flint repeats and strengthens the chronology of the much larger collection of named projectile points from Sudden Shelter (Fig. 17). There, also with good chronological controls, it was possible to outline, with some precision, shifts in projectile point types which largely defined the three components at the Sudden site (Jennings et al. In press). The fully supportive Cowboy Cave evidence, combined with that from Sudden Shelter, allowed Schroedl [1976] to reexamine the scattered Archaic sites reported from the upper Colorado Plateau region using the flint chronological charts as a yardstick. Thus the two sites, Cowboy and Sudden Shelter, provided a standard which permitted the chronological ordering of the Archaic sites in the upper Colorado Plateau into four phases. The Schroedl classification is:

Black Knoll Phase
    Date range: 8300 to 6200 B.P.
    Dominant projectile point types: Pinto points (wherever hunting is a major activity), with Northern Side-notched points occurring in the later half of the phase.

Castle Valley Phase
    Date range: 6200 to 4500 B.P.
    Dominant projectile point types: Several new types named during the analysis of Sudden Shelter, such as the Hawken Side-notched, Sudden Side-notched, and the Rocker Base side-notched points.

Green River Phase
    Date range: 4500 to 3300 B.P.
    Dominant projectile point types: Gypsum points.

Dirty Devil Phase
    Date range: 3300 to 1500 B.P.
    Dominant projectile point types: The Gypsum point in the earlier half, replaced by the Rose Springs arrowpoint toward the end of the phase.

    Figure 18 shows the correlation between the Cowboy Cave strata and the Archaic phases proposed by Schroedl. Units II and III compose the earliest cultural component, which is equivalent to the Black Knoll Phase. During that time period (8300 to 6200 B.P.) the diagnostic projectile point type is the Northern side-notched, although the Elko series is present. Because of the lack of mammal bone recovered and the absence of the Pinto point believed to be associated with artiodactyl hunting [Holmer 1978], the Northern Side-notched and the Elko series might be interpreted as general purpose knives, as mentioned earlier.

    The Castle Valley Phase appears not to be represented at Cowboy Cave by any cultural remains. The next cultural Unit (IV) correlates with the Green River and the Dirty Devil Phases, beginning at the earliest at 4600 B.P. and lasting until ca. 1600 B.P. The diagnostic point is the Gypsum, which appears to have been used for the hunting of bighorn sheep.

    Unit V begins late in the Dirty Devil Phase and is marked by the introduction of Rose Springs points and the bow and arrow at about 1600 B.P. The unit ends at an unknown time after 1500 B.P. The Elko series points continue throughout the Cowboy Cave sequence, supporting the implication that these forms might be knives rather than projectile points. The lack of a specific time niche for the Elko series constitutes an advance in the understanding of the chronology of the Colorado Plateau.

    Little that is convincing can be said about climatic change in the Cowboy Cave vicinity. In *Appendix X*, Lindsay, if I read him right, indicates his belief that there was climatic variation, but it was most marked during periods of disuse. Obviously, there were initially great changes in the environment between the use of the cave by the now-extinct megafauna and man's first use of the cave. This, however, has to do with effects of Pleistocene termination consisting largely of the marked change in the elevation of many coniferous species. After the disappearance of several conifers, when modern vegetation took over, there is less evidence for significant climatic change visible at Cowboy Cave, except in the pollen record. The disappearance of *Corispermum* could represent a climatic change, although it is imputed here to historical overgrazing, which is a very short-term explanation.

    One of the most mysterious and unexplained aspects of the Cowboy Cave situation is the lengthy breaks between occupancy after the cave began to be used by humans. There appears to be little doubt that all the human occupations were short lived, although one lasted 800 years, according to the radiocarbon dates.

    In sum, then, Cowboy Cave intrinsically can be interpreted as a special use, summer season station with a single major activity, i.e., the gathering of economic grains. The paragraphs in this section have attempted to indicate many of the data of both local and regional significance that were gleaned from the evidence. Hunting was at a minimum, and

occupancy was not continuous. Taken
with Sudden Shelter, it would appear to
be one of the most informative of the
Archaic sites now on record in the Colorado Plateau province.

REFERENCES

ADOVASIO, JAMES M.

1970a   The Origin, Development, and
         Distribution of Western Archaic
         Textiles. Ph.D. Dissertation,
         University of Utah Department
         of Anthropology, Salt Lake City.

1970b   Textiles. *In* "Hogup Cave," C.
         Melvin Aikens. *University of
         Utah Anthropological Papers,*
         No. 93. Salt Lake City.

1970c   The Origin and Development of
         Western Archaic Textiles.
         Tebiwa, Vol. 13, No. 2, pp. 1-
         40. Pocatello.

AIKENS, C. MELVIN

1970   Hogup Cave. *University of Utah
         Anthropological Papers,* No. 93.
         Salt Lake City.

AMBLER, J. RICHARD

1966   Caldwell Village. *University of
         Utah Anthropological Papers,*
         No. 84. Salt Lake City.

AMSDEN, CHARLES AVERY

1949   *Prehistoric Southwesterners
         from Basketmaker to Pueblo.*
         Southwest Museum, Los Angeles.

BEETLE, ALAN A.

1970   Recommended Plant Names. *Re-
         search Journal* 31. Agricultural
         Experiment Station, University
         of Wyoming, Laramie.

BETTINGER, ROBERT L.

1976   The Development of Pinyon Ex-
         ploitation in Central Eastern
         California. *Journal of Califor-
         nia Anthropology,* Vol. 3, No.
         1, pp. 81-95. Banning, Cali-
         fornia.

1977   Aboriginal Human Ecology in
         Owens Valley: Prehistoric
         Change in the Great Basin.
         American Antiquity, Vol. 42,
         No. 1, pp. 3-17. Washington,
         D. C.

BOND, D. CLIFFORD

1941   Magic of the Navajo Medicine
         Man. *Arizona Highways,* Vol. 17,
         No. 7. Prescott.

BROWN, F. MARTIN

1942   The Microscopy of Mammalian
         Hair for Anthropologists. *Pro-
         ceedings of the American Philo-
         sophical Society,* Vol. 85, No.
         3. Philadelphia.

BURGH, ROBERT F., AND CHARLES R. SCOGGIN

1948   The Archeology of Castle Park,
         Dinosaur National Monument.

University of Colorado Studies, Series in Anthropology, No. 2. Boulder.

COSGROVE, C. BURTON

1947   Caves of the Upper Gila and Hueco Areas in New Mexico and Texas. *Papers of the Peabody Museum of American Archaeology and Ethnology*, Vol. 24, No. 2, pp. 1-181. Cambridge.

COTTAM, WALTER P.

1947   Is Utah Sahara Bound? *Bulletin of the University of Utah*, Vol. 37, No. 11. Salt Lake City.

CRESSMAN, LUTHER S.

1942   Archaeological Researches in the Northern Great Basin. *Carnegie Institution of Washington Publication*, No. 538. Washington, D. C.

DALLEY, GARDINER F.

1970   Artifacts of Wood. *In* "Hogup Cave," C. Melvin Aikens. *University of Utah Anthropological Papers*, No. 93. Salt Lake City.

1978   Swallow Shelter and Associated Sites. *University of Utah Anthropological Papers*, No. 96. Salt Lake City.

DURRANT, STEPHEN D.

1952   Mammals of Utah, Taxonomy and Distribution. *University of Kansas Publications*, Vol. 6, Museum of Natural History, Lawrence.

ELSASSER, ALBERT B., AND E. R. PRINCE

1961   Eastgate Cave. *University of California Anthropological Records*, Vol. 20, No. 4, pp. 139-49. Berkeley.

FOWLER, DON D.

1963   1961 Excavations, Harris Wash, Utah. *University of Utah An-*

thropological Papers, No. 64, *Glen Canyon Series*, No. 19. Salt Lake City.

(Ed.)
1973   S. M. Wheeler, Archeology of Etna Cave, Nevada: A Reprint. *Desert Research Institute Publications in the Social Sciences*, No. 7. Reno.

FOWLER, DON D., DAVID B. MADSEN, AND EUGENE HATTORI

1973   Prehistory of Southeastern Nevada. *Desert Research Institute Publications in the Social Sciences*, No. 6. Reno.

GILBERT, B. MILES

1973   *Mammalian Osteo-Archaeology: North America*. Special Publications, Missouri Archeological Society, Columbia, Missouri.

GROSSCUP, GORDON L.

1960   The Culture History of Lovelock Cave, Nevada. *Reports of the University of California Archeological Survey*, No. 52. Berkeley.

GRUHN, RUTH

1961   The Archaeology of Wilson Butte Cave, South-Central Idaho. *Occasional Papers of the Idaho State College Museum*, No. 6. Pocatello.

GUERNSEY, SAMUEL J., AND ALFRED V. KIDDER

1921   Basket-Maker Caves of Northeastern Arizona: Report of the Explorations, 1916-17. *Papers of the Peabody Museum of American Archaeology and Ethnology*, Vol. 8, No. 2. Cambridge.

GUNNERSON, JAMES H.

1959   1957 Excavations, Glen Canyon Area. *University of Utah Anthropological Papers*, No. 43, *Glen Canyon Series*, No. 10. Salt Lake City.

1969   The Fremont Culture: A Study in Culture Dynamics on the Northern Anasazi Frontier. *Papers of the Peabody Museum of Archaeology and Ethnology*, Vol. 59, No. 2. Cambridge.

HARRINGTON, H. D.

1967   *Edible Native Plants of the Rocky Mountains.* University of New Mexico Press, Albuquerque.

HARRINGTON, MARK R.

1933   Gypsum Cave, Nevada. *Southwest Museum Papers*, No. 8. Los Angeles.

HAURY, EMIL W.

1950   *The Stratigraphy and Archeology of Ventana Cave, Arizona.* University of Arizona and University of New Mexico Presses, Tucson and Albuquerque. Second Printing, 1975. University of Arizona Press, Tucson.

HEIZER, ROBERT F., AND R. BERGER

1970   Radiocarbon Age of the Gypsum Culture. *University of California Archaeological Research Facility Contributions*, No. 7, pp. 13-18. Berkeley.

HEIZER, ROBERT F., AND ALEX D. KRIEGER

1956   The Archaeology of Humboldt Cave, Churchill County, Nevada. *University of California Publications in American Archaeology and Ethnology*, Vol. 47, No. 1. Berkeley.

HESTER, THOMAS R., AND ROBERT F. HEIZER

1973   *Review and Discussion of Great Basin Projectile Points: Form and Chronology.* Archaeological Research Facility, University of California, Berkeley.

HESTER, THOMAS, M. P. MILDNER, AND L. SPENCER

1974   Great Basin Atlatl Studies.

Ballena Press Publications in Archaeology, Ethnology, and History, No. 2. Ramona, California.

HOGAN, PATRICK

1976   Lithic Analysis. *In* "Sudden Shelter," Jesse D. Jennings, Alan R. Schroedl, and Richard N. Holmer. *University of Utah Anthropological Papers.* Salt Lake City. (In press)

1978   The Analysis of Human Coprolites From Cowboy Cave. Manuscript on file, University of Utah Department of Anthropology. Salt Lake City.

HOLMER, RICHARD N.

1976a   Projectile Points. *In* "Sudden Shelter," Jesse D. Jennings, Alan R. Schroedl, and Richard N. Holmer. *University of Utah Anthropological Papers.* Salt Lake City. (In press)

1976b   Factor Analysis. *In* "Sudden Shelter," Jesse D. Jennings, Alan R. Schroedl, and Richard N. Holmer. *University of Utah Anthropological Papers.* Salt Lake City. (In press)

1977   *A Statistical Analysis of Great Basin Projectile Point Morphology and the Chronological Implications.* Ph.D. dissertation, University of Utah Department of Anthropology. Salt Lake City.

HOWARD, RICHARD M.

1968   *The Mesa Verde Museum.* Library of Congress Number 68-28763. K C Publications, Flagstaff.

HUNT, CHARLES B.

1956   Cenozoic Geology of the Colorado Plateau. *Geological Survey Professional Paper* No. 279, pp. 1-99. U. S. Geological Survey, Government Printing Office, Washington, D. C.

HURST, C. T.

1944   1943 Excavation in Cave II, Tabequache Canyon, Montrose County, Colorado. *Southwestern Lore*, Vol. 10, No. 1, pp. 2-14. Boulder.

1947   Excavations of Dolores Cave--1946. *Southwestern Lore*, Vol. 13, No. 1, pp. 8-17. Boulder.

1948   The Cottonwood Expedition 1947--a Cave and a Pueblo Site. *Southwestern Lore*, Vol. 14, No. 1, pp. 4-19. Boulder.

HYDE, GEORGE E.

1959   *Indians of the High Plains, from the Prehistoric Period to the Coming of the Europeans.* University of Oklahoma Press, Norman.

JENNINGS, JESSE D.

1957   Danger Cave. *University of Utah Anthropological Papers*, No. 27. Salt Lake City.

1966   Glen Canyon: A Summary. *University of Utah Anthropological Papers*, No. 81, *Glen Canyon Series*, No. 31. Salt Lake City.

1968   *Prehistory of North America.* McGraw-Hill, New York.

1975   Preliminary Report: Excavation of Cowboy Caves. Report on file, University of Utah Department of Anthropology. Salt Lake City.

JENNINGS, JESSE D., ET AL.

1976   Sudden Shelter. *University of Utah Anthropological Papers.* Salt Lake City. (In press)

JETT, STEPHEN C.

1968   Grand Canyon Dams, Split-twig Figurines, and "Hit-and-run" Archeology. *American Antiquity*, Vol. 33, No. 3, pp. 341-51. Washington, D. C.

JUDD, NEIL M.

1926   Archaeological Observations North of the Rio Colorado. *Bureau of American Ethnology Bulletin*, No. 82. Washington, D. C.

KEARNEY, THOMAS H., AND ROBERT H. PEEBLES

1964   *Arizona Flora.* University of California Press, Berkeley.

KELLY, ROGER E.

1966   Split-twig Figurines from Sycamore Canyon, Central Arizona. *Plateau*, Vol. 38, No. 3, pp. 65-67. Flagstaff.

KIDDER, ALFRED, AND SAMUEL J. GUERNSEY

1919   Archeological Explorations in Northeastern Arizona. *Bureau of American Ethnology*, Bul. 65. Washington, D. C.

LAMBERT, MARJORIE F., AND RICHARD J. AMBLER

1961   A Survey and Excavation of Caves in Hidalgo County, New Mexico. *The School of American Research Monograph* No. 25. Santa Fe.

LAWRENCE, BARBARA

1951   Post-cranial Skeletal Characters of Deer, Pronghorn, and Sheep-goat with Notes on Bos and Bison. *Papers of the Peabody Museum of American Archaeology and Ethnology*, Vol. 35, No. 3. Cambridge. Reprint 1968 by Kraus Reprint Corporation, New York.

LAYTON, THOMAS N.

1972   Lithic Chronology in the Fort Rock Valley, Oregon. *Tebiwa*, Vol. 15, No. 2, pp. 1-21. Pocatello.

LINDSAY, ALEXANDER J., JR., ET AL.

1968   Survey and Excavations North and East of Navajo Mountain

1959-1962. *Bulletin of the Museum of Northern Arizona*, No. 45, *Glen Canyon Series*, No. 8. Flagstaff.

LIPE, WILLIAM D.

1960   1958 Excavations, Glen Canyon Area. *University of Utah Anthropological Papers*, No. 44, *Glen Canyon Series*, No. 11. Salt Lake City.

LISTER, ROBERT H., ET AL.

1960   The Coombs Site, Part II. *University of Utah Anthropological Papers*, No. 41, *Glen Canyon Series*, No. 8. Salt Lake City.

LOUD, L. L. AND MARK R. HARRINGTON

1929   Lovelock Cave. *University of California Publications in American Archaeology and Ethnology*, Vol. 25. Berkeley.

LUCIUS, WILLIAM A. (ED.)

1976   Archeological Investigations in the Maze District, Canyonlands National Park, Utah. *Antiquities Section Selected Papers*, Vol. 3, No. 11. Salt Lake City.

MADSEN, DAVID B.

1973   Late Quaternary Paleoecology in the Southeastern Great Basin. Ph.D. Dissertation, University of Missouri Department of Anthropology. Columbia.

MARTIN, PAUL S., ET AL.

1952   Mogollon Cultural Continuity and Change: A Stratigraphic Analysis of Tularosa and Cordova Caves. *Fieldiana: Anthropology*, Vol. 40. Chicago.

MARWITT, JOHN P.

1971   Median Village and Fremont Culture Regional Variation. *University of Utah Anthropological Papers*, No. 95. Salt Lake City.

MASON, OTIS TUFTON

1902   Aboriginal American Basketry: Studies in Textile Art Without Machinery. *Annual Report of the U. S. National Museum*. Washington, D. C.

MOORE, TOMMY D., ET AL.

1974   Identification of the Dorsal Guard Hairs of Some Mammals of Wyoming. *Wyoming Game and Fish Department Bulletin* No. 14. Cheyenne.

MORRIS, EARL H., AND ROBERT F. BURGH

1941   Anasazi Basketry, Basket Maker II through Pueblo III, a Study Based on Specimens from the San Juan River Country. *Carnegie Institution of Washington Publication*, No. 533. Washington, D. C.

MORSS, NOEL M.

1931   The Ancient Culture of the Fremont River in Utah. *Papers of the Peabody Museum of American Archaeology and Ethnology*, Vol. 12, No. 3. Cambridge.

1954   Clay Figurines of the American Southwest. *Papers of the Peabody Museum of American Archaeology and Ethnology*, Vol. 49, No. 1. Cambridge.

MULLOY, WILLIAM

1958   A Preliminary Historical Outline for the Northwestern Plains. *University of Wyoming Publications*, Vol. 22, No. 1. Laramie.

MUNSELL SOIL COLOR CHARTS

1975   Munsell Color, Macbeth Division, Kollmorgen Corp. Baltimore.

MUTO, GUY ROGER

1971   A Technological Analysis of the Early Stages in the Manufacture of Lithic Artifacts. M.A. thesis, Idaho State University Department of Anthropology, Pocatello.

NEWCOMB, WILLIAM W., JR.

1976   Pecos River Pictographs. **In**
        Cultural Change and Continuity:
        Essays in Honor of James Ben-
        nett Griffin. Academic Press,
        New York.

NIE, NORMAN H., ET AL.

1975   *SPSS: Statistical Package for*
        the Social Sciences. Second
        edition. McGraw-Hill, New York.

NUSBAUM, JESSE L.

1922   A Basket-Maker Cave in Kane
        County, Utah: With Notes on
        the Artifacts by A. V. Kidder
        and S. J. Guernsey. *Museum of*
        the American Indian, Heye
        Foundation, Indian Notes and
        Monographs, Miscellaneous, No.
        29. New York.

PARSONS, M. L.

n.d.   Painted Pebbles: A Stylistic
        and Chronological Analysis.
        Report on file, Texas Memorial
        Museum, Austin.

PENDERGAST, DAVID M.

1961   1960 Test Excavations in the
        Plainfield Reservoir Area.
        Addendum to *University of Utah*
        Anthropological Papers, No. 52,
        Glen Canyon Series, No. 14.
        Salt Lake City.

PIERSON, LLOYD, AND KEVIN ANDERSON

1975   Another Split-twig Figurine
        from Moab, Utah. *Plateau,* Vol.
        48, Nos. 1-2, pp. 43-45. Flag-
        staff.

PRICE, SARA SUE (RUDY)

1957   Textiles. *In* "Danger Cave,"
        Jesse D. Jennings. *University*
        of Utah Anthropological Papers,
        No. 27, pp. 235-64. Salt Lake
        City.

PURDY, BARBARA ANN

1974   Investigations Concerning the
        Thermal Alteration of Silica
        Materials: An Archaeological
        Approach. *Tebiwa,* Vol. 17, No.
        1, pp. 37-66. Pocatello.

1975   Fractures for the Archaeologist.
        In *Lithic Technology,* Earl
        Swanson (Ed.), pp. 131-41.
        Mouton and Co., The Hague.

REILLY, P. T.

1966   The Sites at Vasey's Paradise.
        Masterkey, Vol. 40, No. 4, pp.
        126-39. Los Angeles.

RIDDELL, FRANCIS A.

1960   The Archaeology of the Karlo
        Site, California. *Reports of*
        the University of California
        Archaeological Survey, No. 53.
        Berkeley.

ROHN, ARTHUR H.

1971   *Wetherill Mesa Excavations, Mug*
        House. Mesa Verde National Park,
        Colorado.

RUDY, JACK R.

1953   Archeological Survey of Western
        Utah. *University of Utah Anthro-*
        pological Papers, No. 12. Salt
        Lake City.

RYDBERG, PER AXEL

1929   *Flora of the Rocky Mountains*
        and Adjacent Plains of Colorado,
        Utah, Wyoming, Idaho, Montana,
        Saskatchewan, Alberta, and
        Neighboring Parts of Nebraska,
        South Dakota, North Dakota, and
        British Columbia. Published by
        the author for the New York
        Botanical Garden. New York.

SCHROEDER, ALBERT H.

1955   Archeology of Zion Park. *Univer-*
        sity of Utah Anthropological
        Papers, No. 22. Salt Lake City.

SCHROEDL, ALAN R.

1976   The Archaic of the Northern
Colorado Plateau. Ph.D. dissertation, University of Utah Department of Anthropology, Salt
Lake City.

1977   The Grand Canyon Figurine Complex. *American Antiquity*, Vol.
42, No. 2, pp. 254-65. Washington, D. C.

SCHUSTER, CARL

1968   Incised Stones from Nevada and
Elsewhere. *Nevada Archeological
Survey Reporter*, Vol. 2, No. 5,
pp. 4-23. Reno.

SCHWARTZ, DOUGLAS W., ET AL.

1958   Split-twig Figurines in the
Grand Canyon. *American Antiquity*, Vol. 23, No. 3, pp. 264-
74. Washington, D. C.

SHAFER, HARRY J.

1975   Clay Figurines From the Lower
Pecos Region, Texas. *American
Antiquity*, Vol. 40, No. 2, pp.
148-58. Washington, D. C.

SHARROCK, FLOYD W., ET AL.

1961   1960 Excavations, Glen Canyon
Area. *University of Utah An-
thropological Papers*, No. 52,
Glen Canyon Series, No. 14.
Salt Lake City.

1963   1961 Excavations, Glen Canyon
Area. *University of Utah An-
thropological Papers*, No. 63,
Glen Canyon Series, No. 18.
Salt Lake City.

SHARROCK, FLOYD W., AND EDWARD G. KEANE

1962   Carnegie Museum Collection from
Southeastern Utah. *University
of Utah Anthropological Papers*,
No. 57, *Glen Canyon Series*, No.
16. Salt Lake City.

SHELLBACH, LOUIS

1922   Ancient Bundles of Snares from

Nevada. *Museum of the American
Indian Heye Foundation, Indian
Notes and Monographs*, Vol. 4,
No. 3, pp. 232-40. New York.

SHUTLER, RICHARD, JR.

1967   Cultural Chronology in Southern
Nevada. *In* "Pleistocene Studies
in Southern Nevada, Part 6," H.
Marie Wormington and D. Ellis
(Eds.), *Nevada State Museum Anthropological Papers*, No. 13,
pp. 305-08. Reno.

SPIER, LESLIE

1958   Mohave Culture Items. *Museum of
Northern Arizona Bulletin*, No.
28. Flagstaff.

STEWARD, JULIAN H.

1937   Ancient Caves of the Great Salt
Lake Region. *Bureau of American
Ethnology Bulletin*, No. 116.
Washington, D. C.

1938   Basin-Plateau Aboriginal Socio-
political Groups. *Bureau of
American Ethnology Bulletin*,
No. 120. Washington, D. C. Re-
print 1970 by University of Utah
Press, Salt Lake City.

STREUVER, STUART

1968   Flotation Techniques for the
Recovery of Small Scale Archaeological Remains. *American Antiquity*, Vol. 33, No. 3, pp. 353-
62. Washington, D. C.

SWANSON, EARL H., ET AL.

1964   Birch Creek Papers No. 2: Natural and Cultural Stratigraphy in
the Birch Creek Valley of Eastern Idaho. *Occasional Papers of
the Idaho State University
Museum*, No. 14. Pocatello.

SWEET, MURIAL

1962   *Common Edible and Useful Plants
of the West*. Naturegraph Company,
Healdsburg, California.

TAYLOR, DEE C.

1957   Two Fremont Sites and Their
        Position in Southwestern Pre-
        history. *University of Utah
        Anthropological Papers*, No. 29.
        Salt Lake City.

TREVOR-DEUTSCH, BURLEIGH

1970   Hair Morphology and Its Use in
        the Identification of Taxonomic
        Groups. Ms. on file, Department
        of Plant Pathology, MacDonald
        College, McGill University,
        Montreal.

TRIPP, GEORGE W.

1967   An Unusual Split-willow Figu-
        rine Found near Green River,
        Utah. *Utah Archeology*, Vol. 13,
        No. 1, p. 15. Salt Lake City.

TURNER, CHRISTY G., II

1962   A Summary of the Archeological
        Explorations of Dr. Byron Cum-
        mings in the Anasazi Culture
        Area. *Museum of Northern Ari-
        zona Technical Series*, No. 5.
        Flagstaff.

UBELAKER, DOUGLAS H., AND WALDO R. WEDEL

1975   Bird Bones, Burials, and Bun-
        dles in Plains Archeology.
        American Antiquity, Vol. 40,
        No. 4, pp. 444-452. Washington,
        D. C.

UNDERHILL, RUTH

1944   *Pueblo Crafts*. U. S. Department
        of the Interior Bureau of
        Indian Affairs. Washington,
        D. C.

WELTFISH, GENE

1930   Prehistoric North American
        Basketry Techniques and Modern
        Distributions. *American Anthro-
        pologist*, Vol. 32. Washington,
        D. C.

1932   Problems in the Study of
        Ancient and Modern Basket-
        Makers. *American Anthropologist*,
        Vol. 34, No. 1. Washington,
        D. C.

WHEAT, MARGARET M.

1967   *Survival Arts of the Primative
        Piutes*. University of Nevada
        Press. Reno.

WHEELER, S. M.

1942   *Archeology of Etna Cave, Lincoln
        County, Nevada*. Nevada State
        Park Commission, Carson City.

WORMINGTON, H. MARIE

1966   *Prehistoric Indians of the
        Southwest*. Denver Museum of
        Natural History, Denver.

WORMINGTON, H. MARIE, AND ROBERT H.
 LISTER

1956   Archaeological Investigations
        on the Uncompahgre Plateau in
        West Central Colorado. *Proceed-
        ings of the Denver Museum of
        Natural History*, No. 2. Denver.

WYMAN, LELAND C. AND STUART K. HARRIS

1951   The Ethnobotany of the Kayenta
        Navaho. *University of New Mexico
        Publications in Biology*, No. 5.
        Albuquerque.

YARNELL, RICHARD A.

1965   Implications of Distinctive
        Flora on Pueblo Ruins. *American
        Anthropologist*, Vol. 67, No. 3,
        pp. 662-74. Menasha.

LATE PLEISTOCENE MONTANE CONIFERS IN SOUTHEASTERN UTAH

W. Geoffrey Spaulding and
Thomas R. Van Devender

The macrofossil record of montane conifers in the southwestern United States is poorly known. During the summer of 1975, archeological excavations directed by Jesse D. Jennings, University of Utah, at Cowboy Cave, Wayne County, Utah (110°13'W, 38°19'N; 1,710 m. elevation), exposed Pleistocene-age deposits beneath the oldest cultural unit. One stratum, approximately 50 cm. thick, is composed almost entirely of herbivore dung. Mixed with the dung are needles of spruce (*Picea engelmanni* or *P. pungens*) and Douglas fir (*Pseudotsuga menziesii*). A radiocarbon date on dung from the entire 50 cm. was 11,810±40 B.P.(before present; UGa636).

The present plant community near Cowboy Cave on the Navajo Sandstone is a well developed pinyon-juniper woodland dominated by Colorado pinyon (*Pinus edulis*) and Utah juniper (*Juniperus osteosperma*). Other important associates are Gambel oak (*Quercus gambelii*), buffaloberry (*Shepherdia rotundifolia*) and serviceberry (*Amelanchier utahensis*). Big sagebrush (*Artemisia tridentata*) communities are common on the sandy flats between the canyons. Relict stands of Douglas fir presently occur in canyons within 10 km. of Cowboy Cave at elevations as low as 1,980 m. The nearest spruce populations occur at elevations as low as 2,140 m. in mesic situations in the Henry Mountains, 50 km. to the southwest [Everitt 1970].

The Pleistocene record of spruce and Douglas fir at 1,710 m. elevation in southeastern Utah is of considerable interest in light of the previous records of montane conifers. Mehringer and Ferguson (1969) found intermountain bristlecone pine (*Pinus longaeva*), limber pine (*P. fiexilis*), and white fir (*Abies concolor*) in fossil packrat middens at 1,910 m. elevation on Clark Mountain, San Bernardino County, California. Radiocarbon dates on the middens were 23,600± 950 B.P. (1-3557) and 28,720±1800 B.P. (1-3648). Similar assemblages were reported from the Sheep Range, Clark County, Nevada, by Spaulding (1974). Madsen (1972, [1973]) reported bristlecone and limber pine from packrat middens in Meadow Valley Wash, Lincoln County, Nevada as low as 1,340 m. This bristlecone pine record is unusual because of the low elevation and because it is on Tertiary volcanic substrates (bristlecone pine is generally restricted to limestone or dolomite). None of the Pleistocene records of montane conifers in the Mohave Desert and adjacent areas contain spruce or Douglas fir. Today these two species reach their western limit in east central Nevada and western Utah. They are absent from most of the Great Basin montane islands (Little 1971). The Cowboy Cave area, in the Glen Canyon drainage of the Colorado River, is a portion of the Colorado Plateau.

The only previous Pleistocene record

of spruce and Douglas fir in the Southwest is from the Guadalupe Mountains, Culberson County, Texas (Van Devender et al. 1976 In press). The fossils were from a packrat midden and cave fill at 2,000 m. elevation on limestone. Radiocarbon dates on spruce needles were 13,060±280 (A-1549) and 13,000±730 B.P. (A-1539). Associated with spruce and Douglas fir in the Guadalupe Mountains assemblages were dwarf juniper (*Juniperus communis*) and southwestern limber pine (*Pinus flexilis*). White fir was not present.

Though the edaphic influence of sandstone substrate may account for the lower elevation of montane vegetation at Cowboy Cave, it is, nevertheless, the lowest Pleistocene-age record of spruce to date. Wells and Berger (1967) found that late Pleistocene woodlands occurred lower on sandstone than on limestone in southern Nevada. The Cowboy Cave fossils suggest that some of the montane islands in south central Utah were probably connected during the last glaciopluvial period. However, it is unlikely that the forest was continuous from Cowboy Cave to the Henry Mountains, since the intervening terrain dips to 1,280 m. elevation along the Dirty Devil River. Montane forest corridors between the Cowboy Cave area and higher mountains east of the Colorado River are even more unlikely, as the elevations along the river are as low as 1,110 m. Distant transport of seeds probably accounts for the dispersal and lack of continuity of montane conifer forests on the Colorado Plateau.

[*Acknowledgments*. We wish to express our gratitude to Jesse D. Jennings, Department of Anthropology, University of Utah, for his aid and cooperation. Beverly Albee of the University of Utah and Ranger James Walters, National Park Service, were of great help. Research is being conducted with the aid of NSF grant DEB 75-13944 to Paul S. Martin. Betty Fink aided in typing and editing. This is Contribution number 10, Department of Geosciences, University of Arizona, Tucson.]

Addendum: Since the writing of this manuscript, and at the request of David Madsen, the fossil material from Meadow Valley Wash, Nevada was reexamined. Late Pleistocene macrofossils previously reported as being those of bristlecone pine (*Pinus aristata* Engelm.) and white fir (*Abies concolor* Lindl.) are, in fact, limber pine (*Pinus flexilis* James) and Douglas fir (*Pseudotsuga menziesii* Franco). The regional plant association of the Meadow Valley Wash is anomalously mesic, with Gambel oak (*Quercus gambelii* Nutt.), single-leaf pinyon (*Pinus monophylla* Torr. & Frém.), and Utah juniper (*Juniperus osteosperma* Little) occurring as much as 300 m. lower here than in adjacent mountain ranges. Consequently, estimates of plant zone displacement during the Pleistocene are much less than had previously been claimed (Madsen 1972, 1973).

The record of Douglas fir from Meadow Valley Wash is of considerable biogeographic importance. It documents a westward range extension of this species by 100 km. from its most westerly position today (Little 1971). Douglas fir is currently absent from most of the Great Basin region; its westerly expansion during late Wisconsin times may indicate increased summer precipitation. A revision of the paleobotanical data from Meadow Valley Wash is being prepared for publication at this time [Spaulding and Madsen, in prep.].

REFERENCES

EVERITT, BENJAMIN L.

1970   A Survey of the Desert Vegetation of the Northern Henry Mountains Region, Hanksville, Utah. Ph.D. dissertation, John Hopkins University, Baltimore.

LITTLE, ELBERT L., JR.

1971   Atlas of United States Trees, Vol. 1, Conifers and Important Hardwoods. *U. S. Department of Agriculture Miscellaneous Publications*, No. 1146. Washington, D. C.

MADSEN, DAVID B.

1972 Paleoecological Investigations in Meadow Valley Wash, Nevada. *In* "Great Basin Cultural Ecology, A Symposium," Don D. Fowler (Ed.). *Desert Research Institute Publications in the Social Sciences*, No. 8, Reno.

1973 Late Quaternary Paleoecology in the Southeastern Great Basin. Ph.D. dissertation, University of Missouri, Columbia.

MEHRINGER, PETER J., JR., AND C. WESLEY FERGUSON

1969 Pluvial Occurrence of Bristlecone Pine (*Pinus aristata*) in a Mohave Desert Mountain Range. *Journal of the Arizona Academy of Science*, Vol. 5, No. 4, pp. 284-92. Tempe.

SPAULDING, W. GEOFFREY

1974 Dynamics of Late Pleistocene Vegetation Change in Southern Nevada. *Bulletin of the Ecological Society of America*, Vol. 55, p. 27. Durham.

SPAULDING, W. GEOFFREY, AND DAVID B. MADSEN

A Revision of the Paleobotanical Data From Meadow Valley Wash, Southern Nevada. (Manuscript in preparation)

VAN DEVENDER, THOMAS R., ET AL.

1976 Late Pleistocene Biotic Communities From the Guadalupe Mountains, Culberson County, Texas. In *Transactions-Symposium on the Biological Resources of the Chihuahuan Desert Region, U. S. and Mexico*, Roland H. Wauer and David Riskind (Eds.). U. S. National Park Service, Washington, D. C.

n.d. Late Pleistocene Plant Communities in the Guadalupe Mountains, Culberson County, Texas. In *Biological Investigations in the Guadalupe Mountains National Park, Texas, Symposium Volume*, Hugh H. Genoways and Robert L. Baker (Eds.). U. S. National Park Service, Washington, D. C. (In press)

WELLS, PHILLIP V., AND RANIER BERGER

1967 Late Pleistocene History of Coniferous Woodland in the Mohave Desert. *Science*, Vol. 155, No. 3770, pp. 1640-47, Washington, D. C.

APPENDIX II
LATE PLEISTOCENE AND EARLY HOLOCENE PALEOECOLOGY OF COWBOY CAVE

W. Geoffrey Spaulding and
Kenneth L. Petersen

Knowledge of North American paleoecology during the Pleistocene-Holocene transition is critical to understanding Paleo-Indian adaptations in the New World. The most dramatic environmental changes prior to the advent of technological man occurred between 14,000 and 8,000 radiocarbon years before the present (B.P.). The wasting of the massive Laurentide and Cordilleran ice sheets, which began abruptly about 14,000 B.P. (Denton 1974, Fig. 13), was accompanied by a trend toward warmer and effectively drier climatic regimes at lower latitudes in North America. Pluvial lakes waned and disappeared. Biotic communities experienced species turnovers. The extinction of most of the North American megafauna also occurred during this period. These climatic and biotic events may have had considerable impact upon prehistoric human populations. Rapid changes from technologies adapted toward big-game hunting to those apparently exploiting a more diversified resource base are obvious in the archeological record. The deposits in Cowboy Cave span most of this critical period; it is the only cave in the western hemisphere known to contain copious quantities of the dung of large Pleistocene herbivores other than the extinct Shasta ground sloth (*Nothrotheriops shastense*). As such, it is a remarkable source of information on the paleoecology of extinct large vertebrates (see Hansen, *Appendix III*). Our

analysis was directed toward reconstructing the local late Pleistocene-early Holocene environment using the plant remains and pollen in Strata Ia, Ib and IIa.

LOCATION AND STRATIGRAPHY

Cowboy Cave is located at 1,710 m. elevation on the south-facing slope of a small canyon. It is a large shelter developed by exfoliation of the basal Navajo Sandstone near its contact with the underlying Kayenta Formation. The more resistant Kayenta Formation forms the bedrock of the wash at the foot of the talus slope below Cowboy Cave.

The cave is situated within the lower limits of the pinyon-juniper woodland. Common species near the mouth of the cave include Gambel oak (*Quercus gambelii*), Utah juniper (*Juniperus osteosperma*), Colorado pinyon (*Pinus edulis*) and Mohave prickly pear cactus (*Opuntia erinacea*). Along the wash below the cave are such shrubs as serviceberry (*Amelanchier utahensis*), buffaloberry (*Shepherdia rotundifolia*), wild rose (*Rosa woodsii*) and singleleaf ash (*Fraxinus anomala*). Mesas between bedrock exposures of canyons and hills support well-developed stands of big sagebrush (*Artemisia tridentata*) on deep, sandy soil. Subdominants of the big sagebrush community include joint-fir (*Ephedra nevadensis*) and Indian ricegrass (*Ory-zopsis hymenoides*). Botanical nomenclature follows Kearney and Peebles (1964).

The lowest stratum, Ia, consists of reasonably well-sorted sands and silts with carbonate inclusions up to 0.5 cm. in dia. The Stratum Ia/Ib contact is gradational over ca. 8 cm. Decomposition of the dung in Ib increases with depth until it takes on the attributes of Stratum Ia. Rodent burrowing has further obscured the Ia/Ib contact. Tunnels were found leading away from the cave wall under Stratum Ib. Rodents (cf. *Neotoma*) apparently used the compacted Ib as a roof, preferring it to the unconsolidated units above. This may be a common phenomenon in caves where indurated and loose strata alternate. It has been attributed to pocket mice (*Perognathus* sp.) in Flaherty Shelter [Spaulding 1974]. Rodent burrowing may have caused contamination in Sandia Cave, New Mexico (Bliss 1940, 1941) and at Gypsum Cave, Nevada (Harrington 1933).

The compacted, uriniferous dung of Stratum Ib contains intercalated lenses of eolian sand. It differs radically from dung strata found in ground sloth caves of the southwest (Martin et al. 1961; Long et al. 1974). The plant fragments in the dung are of a finer texture, compacted, and deposited in "cowpie" form. This contrasts with the coarsely comminuted plant fragments in sloth dung, the unconsolidated nature of sloth dung strata, and the softball-sized bolus of the Shasta ground sloth (see Hansen, *Appendix III*). Trampling appears to have been extensive in Stratum Ib. A hard, compacted floor containing artiodactyl and rabbit fecal pellets marks the top of the stratum. Fecal pellets were not found further down in Stratum Ib. The floor indicates a depositional hiatus between Strata Ib and IIa. This break is shown in Figures 1 and 2.

Stratum IIa differs radically from the strata below it and marks the inception of an eolian depositional environment in Cowboy Cave. It is composed of wind-blown sand and silt with lenses of plant debris. Both the sand and the intercalated plant macrofossils appear to have been deposited near the mouth of the cave by wind eddies.

PALYNOLOGY

Samples for pollen analysis were taken near the mouth of the cave during the 1975 field season. The pollen profile was taken at 4 cm. intervals through Strata Ia, Ib, and IIa. Approximately 100 gm. of matrix were taken for each pollen sample; all samples were placed in sterile plastic bags for transport.

A subsample of approximately 4 cc. of material was taken from each of the 100 gm. pollen samples, dried one hour at 100° C, cooled in a desiccator, and weighed. Eight commercially prepared *Lycopodium* spore tablets (Batch No. 212761), each containing $12,500\pm500$ spores (Stockmarr 1971, 1973) were added to each sample to act as an exotic tracer. Extraction followed Mehringer (1967a) with the addition of acetolyzation. Two samples (13 and 14) were rerun. The second time, approximately 12 cc. of material was used and ten *Eucalyptus* pollen tables, each containing $19,900\pm1750$ grains (L. J. Maher, pers. comm.), were added as a tracer. Samples were placed in vials in a silicon oil medium and slides for pollen counts were prepared from these. A minimum of 330 fossil grains were counted in each sample. In three samples from Strata Ib, chunks of dung and undifferentiated matrix were analyzed separately. Absolute amounts of fossil pollen per gram of sediment were determined by the following formula (Stockmarr 1971, 1973):

$$\frac{(\text{Number of exotics added}) \times (\text{Number of fossil grains counted})}{\text{Number of exotics counted}} =$$

$$\frac{\text{Number of fossil pollen grains}}{\text{Weight of sample in grams}} = \frac{\text{Fossil pollen grains}}{\text{Gram of sediment}}$$

Figure 1 presents the pollen data in terms of relative pollen frequencies (RPF) plotted against depth for Strata Ia, Ib and IIa. Pollen taxa are generalized to generic or family level using

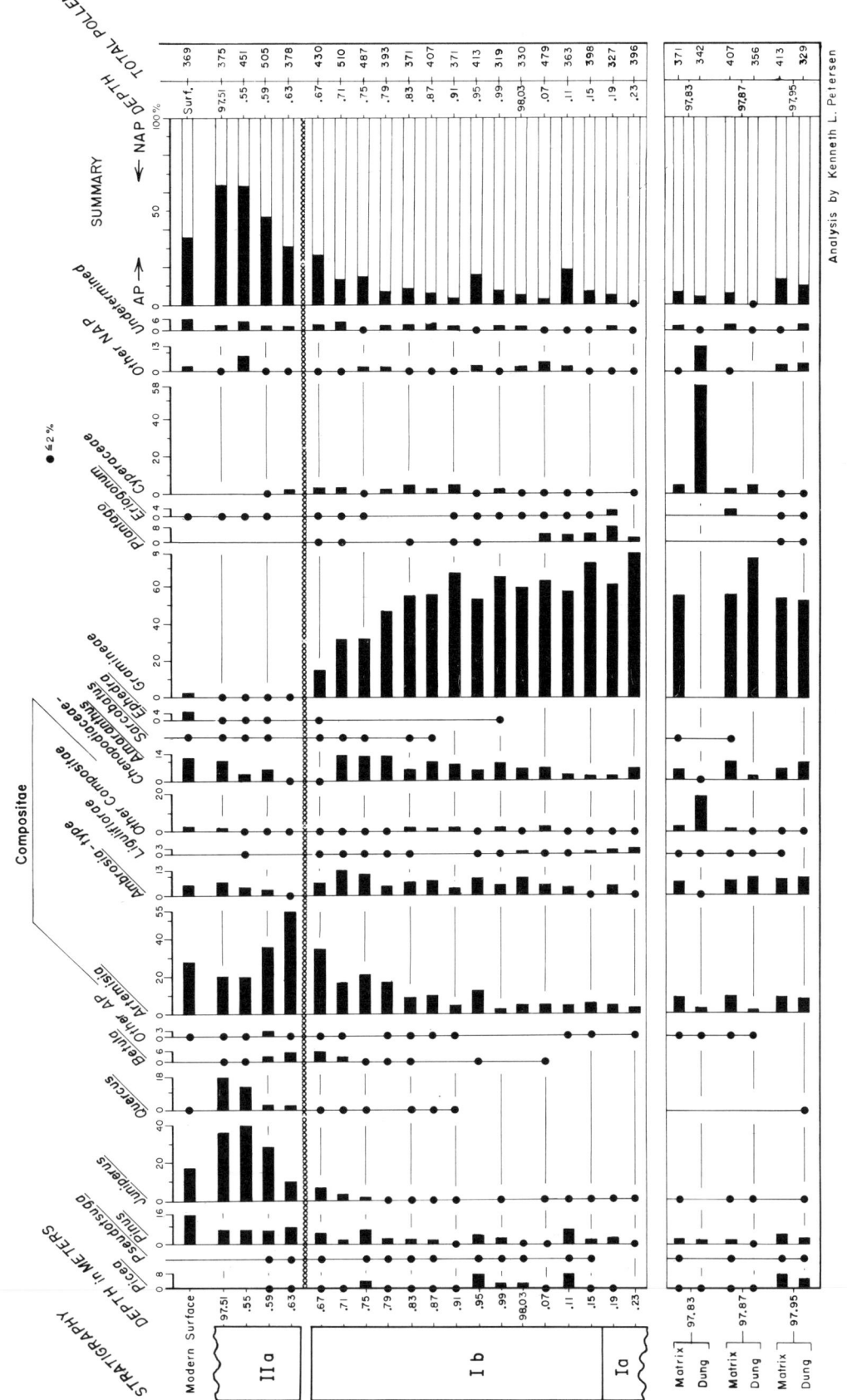

FIG. 1. Pollen percentage diagram.

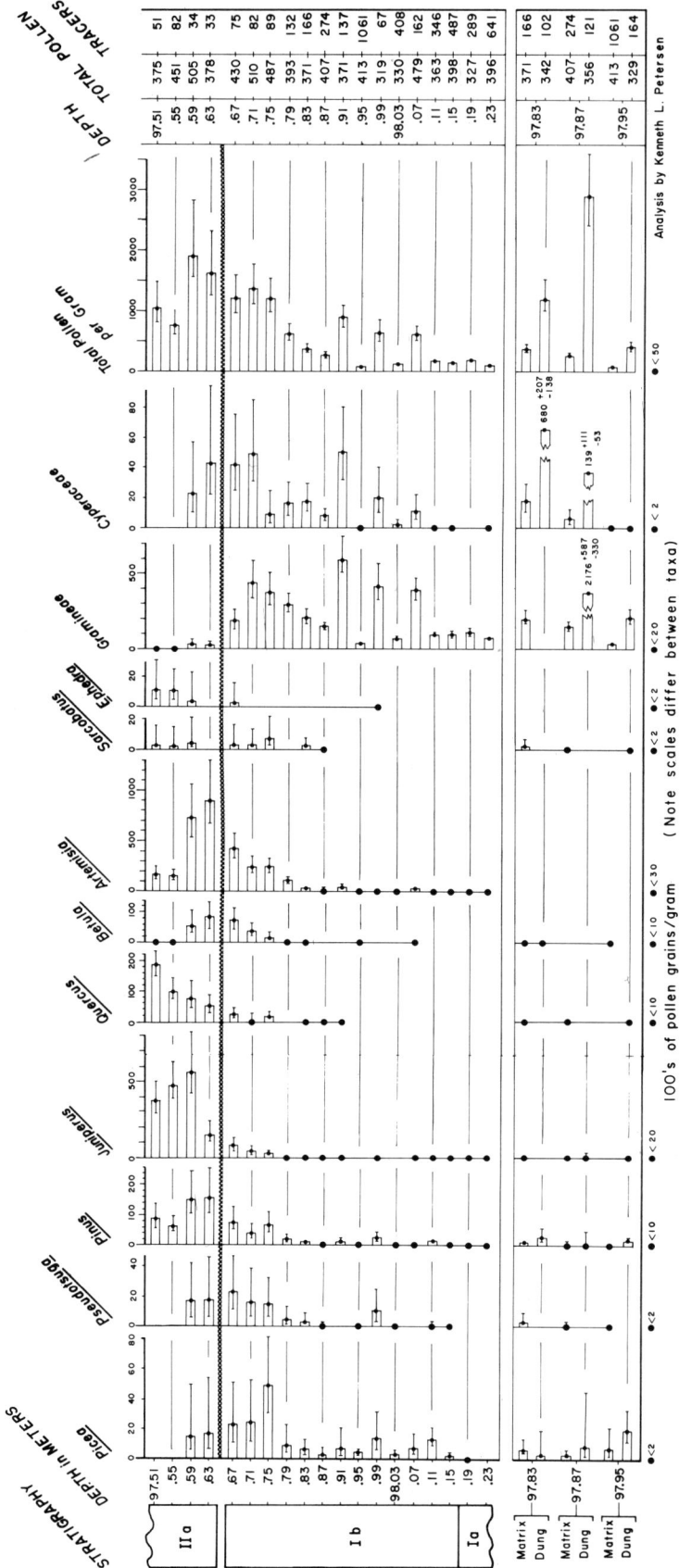

FIG. 2. Absolute pollen diagram of selected taxa.

the terminology of Martin (1963). The RPF of a modern soil sample from a big sagebrush community near the site is shown at the top of Figure 1. Absolute pollen frequencies (APF) for selected taxa with 95 percent confidence levels as determined by Maher's (1972) nomograms are shown in Figure 2. Relative frequencies are under considerable statistical constraint (Davis 1963, Mosimann 1965) and are often difficult to evaluate without accompanying absolute values (Davis et al. 1974). Decrease in the RPF of one pollen type must be accompanied by an increase in at least one other type. For example, the increase of oak (*Quercus*) pollen in the top two samples of Stratum IIa (Fig. 1) is an increase in *Quercus* pollen grains per gram (Fig. 2) rather than a function of a decrease in another pollen type. Three Stratum Ib sample subsets where dung and undifferentiated matrix were analyzed separately are illustrated at the bottom of Figures 1 and 2. Pollen counts are given in Table 1.

Stratum Ia sediments are devoid of pollen, save for the top 8 cm., the transition zone with Stratum Ib (Figs. 1 and 2). Pollen frequencies show little change across the Ia/Ib boundary. Thirty-two pollen taxa were identified from pollen analysis of Stratum Ib. High frequencies of grasses (Gramineae) and sedges (Cyperaceae) reflect the pollen content of herbivore dung, the principal component of this stratum. Cuticle analysis by Hansen (*Appendix III*) shows that dung, tentatively identified as bison, mammoth, and horse, contains more than 95 percent grasses and sedges. Very high APF of grasses and sedges (Fig. 2) in the subsets of dung demonstrate the effect of this dietary bias.

As the amount of pollen per gram of sediment increases towards the top of Stratum Ib, pollen types not introduced by herbivore dung seem to increase. These were probably derived from the local pollen rain, and include *Artemisia* and most arboreal pollen types (AP). Associated with this phenomenon is an increase in the ratio of AP to non-arboreal pollen (NAP; Fig. 1). Significant frequencies of birch pollen (*Betula*) imply the presence of water birch (*Betula fontinalis*) near the cave.

The compositional difference between Strata Ib and IIa is reflected in the pollen profile, particularly in the curve for Gramineae. Twenty-nine pollen taxa were identified by pollen analysis of Stratum IIa sediments. Spruce (*Picea*), Douglas fir (*Pseudotsuga*), birch, and sedge pollen persist into Stratum IIa. The pollen of cottonwood or aspen (*Populus*) and willow (*Salix*) are common in lower Stratum IIa (Table 2). *Populus* pollen, usually poorly preserved in fossil records (Sangster and Dale 1961), is well preserved in the dry environment of Cowboy Cave.

The four samples from Stratum IIa show a dramatic change in the pollen spectra in the top 8 cm. of that stratum. The lower two pollen samples (97.63 m. and 97.59 m., respectively) are characterized by 55 percent and 37 percent *Artemisia* pollen, 3 percent and 4 percent *Quercus* and traces ($\leq$ 2 percent) of *Picea* and *Pseudotsuga* pollen (Fig. 1). The upper two samples (97.55 m. and 97.51 m., respectively) are characterized by 24 percent and 12 percent *Artemisia* pollen, 13 percent and 18 percent *Quercus*, and no *Picea* or *Pseudotsuga* pollen. Single true fir (*Abies*) pollen grains were encountered in the 97.55 m. and 97.63 m. samples (samples 2 and 4, Table 1). *Juniperus* pollen is only 9 percent in the lowermost sample (97.63 m.), and increases to at least 29 percent in the three samples above it in Stratum IIa. Xerophytic elements also increase toward the top of Stratum IIa. These include greasewood (*Sarcobatus*), goosefoot and/or amaranth (Chenopodiaceae-*Amaranthus*) and *Ephedra*. There is also a decrease in the amount of pollen per gram of sediment toward the top of Stratum IIa.

PLANT MACROFOSSILS

Samples for macrofossil analysis of Stratum Ib were taken at 10 cm. intervals from the top of the stratum at grid

TABLE 1

POLLEN COUNTS FROM STRATA Ia, Ib, AND IIa

Stratum	Ia							IIb							IIa								Modern Surface
Sample No.	19	18	17	16	15	14	13	12b	12a	11	10b	10a	9b	9a	8	7	6	5	4	3	2	1	
Depth below datum (cm.)	98.23	98.19	98.15	98.11	98.07	98.03	97.99	97.95 (dung)	97.95	97.91	97.87 (dung)	97.87	97.83 (dung)	97.83	97.79	97.75	97.71	97.67	97.63	97.59	97.55	97.51	
Picea	3	2	7	29	6	10	7	16	33	3	10	5	1	6	6	20	9	8	4	4			
Pseudotsuga			1	2		3	5	3	3			1		3	3	6	6	8	4	4			
Pinus		11	11	29	7	6	12	12	20	5	1	10	8	10	13	28	15	27	37	38	37	31	59
Abies																			1		1		
Juniperus	1	2	3	6	2	2	1	1		1	1	2		3	1	10	16	29	35	144	182	135	62
Quercus								1		1		3		1		6	3	8	12	19	58	67	6
Betula					1				1				1	5	2	6	14	26	19	14	2	1	
Acer													1										
Carya																							1
Alnus												1	1					1	3	1		1	
cf. *Ostrya*														1	1			1		1			
Populus	1	7	6	2						3		3	3	3				5	1	10	7	5	
Salix											1			1	1	1	1	1	1	2	1	2	2
Artemisia	14	17	24	17	25	14	9	28	36	17	10	41	10	36	68	101	91	151	209	185	90	61	102
Ambrosia-type	6	17	6	16	21	27	17	29	34	14	33	32	3	28	21	55	66	29	6	15	18	26	21
Liguliflorae	14	9	8	4	1	7			3		1	4	2	1	1	3	2	1			1		
Cirsium-type				1				1															
Other Compositae	2	4	3	5	13	2	7	4	8	8	2	8	66	9	4	6	3	4	5	4	3	8	10
Chenopodiaceae/*Amaranthus*	21	9	9	10	33	20	29	31	22	32	9	39	2	22	50	65	70	25	13	29	14	25	43
Sarcobatus							1	1				1		2		3	1	1	1	1	3	1	4
Gramineae	309	200	288	204	303	196	207	173	222	250	267	227		205	185	153	163	66					10
Ephedra nevadensis-type																		1			1	1	14
Ephedra trifurca-type							1													1	5	3	2

Taxon	10,000	18,800	13,950	17,350	61,650	11,500	64,050	40,200	7,300	88,400	290,100	26,950	119,700	37,900	62,500	119,850	136,200	120,400	162,100	199,800	76,700	104,400
Rhus																			1			
Platyopuntia																						1
Rosaceae	3	4	3	7	4			3	5			3	1	3		5	3	1	1		3	1
Sphaeralcea-type	2	1		8	1			1	1				2	40						1	1	
Campanula-type					1									1								
Caryophyllaceae	2	1	1	2	1	1		1	1		1		1		1		1	1				1
Eriogonum	12	2	1	7	2	2	3	4	4		2				1		1	1			2	1
Cruciferae								2			12									1		
Phacelia-type																						1
Liliaceae							1	1	1													
Nyctaginaceae									1													
Onagraceae		3		6	3	3	6	3			3		1	3		1						
Gilia-type					1	1	1															
Phlox	2	1		1	3			1	3		1				3	2	2					
Plantago	14	26	18	18	24			1	4	1		1				2	2	2				
Ranunculaceae				2					1	1		1						1	3			
Euphorbia-type																						1
Pedicularis																						2
Polygonum californicum-type										1												
Cyperaceae	1	2	2	9	6	10	1	2	21	17	10	18	197	10	4	19	15	10	6			1
Unknown C																					1	1
Unknown D			1																			
Other unknowns		1												1	1							
Undetermined	5	8	5	7	3	8	9	11	7	8	2	14	5	11	16	9	23	16	9	12	22	11
Tracer:																						
Lycopodium	641	289	487	346	162			164	1061	137	121	274	102	166	132	89	82	75	33	34	82	81
Eucalyptus						408	67															
Pollen/gram dry weight	10,000	18,800	13,950	17,350	61,650	11,500	64,050	40,200	7,300	88,400	290,100	26,950	119,700	37,900	62,500	119,850	136,200	120,400	162,100	199,800	76,700	104,400
N	396	327	398	363	479	330	319	329	413	371	356	407	342	371	393	487	510	430	378	505	451	375

TABLE 2

PLANT MACROFOSSILS FROM STRATA Ib AND IIa

Species	Type	Stratum IIa	Stratum Ib — Depth Below Datum at Grid Square 14R19				
			97.67–97.77	97.77–97.87	97.87–97.97	97.97–98.07	98.07–98.17
Picea cf. *engelmanni*	Leaves	x	x	x	x	x	x
Pseudotsuga menziesii	Leaves	x		x	x		
Pinus edulis	Leaves, Seeds		x	x	x		
Quercus gambelii	Leaves, Seeds	x			x		
Juniperus osteosperma	Twigs, Seeds	x		x	x	x	x
Juniperus scopulorum	Twigs	x		x			
Fraxinus anomala	Leaves	x					
Amelanchier utahensis	Leaves	x					
cf. *Rhamnus smithii*	Leaves, Seeds	x			x		
Populus sp.	Leaves	x					
Cornus stolonifera	Seeds	x					
Rhus trilobata	Seeds	x					
Berberis fremontii	Leaves	x					
Shepherdia rotundifolia	Leaves	x					
Purshia tridentata	Leaves	x					
Rosa woodsii	Leaves	x					
Cercocarpus intricatus	Leaves	x					
Ephedra sp.	Seeds					x	
Artemisia sp.	Bark				x	x	
Opuntia erinacea	Seeds	x					
Opuntia macrorhiza	Seeds	x					
Opuntia sp.	Spines			x	x	x	x
Yucca sp.	Leaves	x					x
Gutierrezia sp.	Leaves, Flowers	x					
Chenopodium sp.	Seeds	x					
Corispermum sp.	Seeds	x					
Eriogonum sp.	Twigs		x				
Stipa sp.	Seeds	x			x		
Oryzopsis hymenoides	Seeds	x					
Agropyron sp.	Seeds	x					
Carex sect. *Orthocerates*	Seeds, Flowers	x					
Carex sp.	Seeds	x					x
Equisetum sp.	Nodes	x		x	x	x	x

square 14R19. Macrofossil samples rep-resenting all of Stratum IIa were taken at grid squares 15R24 and 16R26. For each sample, approximately 5 kg. of matrix were placed in a sterile plastic bag.

Macrofossils were recovered by sift-ing approximately 1 kg. of matrix through a 6 mesh/in. (3.35 mm.) Tyler soil sieve. Larger plant fragments and pieces of dung were set aside for later analysis. Material passing through the 6 mesh sieve was soaked in a detergent solution for three days and then washed through a 20 mesh (0.33 mm.) sieve under running water. Material caught on the 20 mesh sieve was dried and sorted for identifiable fossils. The plant macro-fossils are given in Table 2.

Fifteen plant taxa were identified in the macrofossil analysis of Strata Ib. Some appear to have been introduced by decomposing dung, some by eolian or rodent transport. Macrofossils from the herbivore dung include the aquatics, sedge (*Carex*, Cyperaceae) and horsetail (*Equisetum*, Equisetaceae), grasses, sagebrush, and bark and spines of prickly pear or cholla (*Opuntia* sp.). The most notable macrofossils are the needles of spruce and Douglas fir. Twigs of Rocky Mountain juniper (*Juni-perus scopulorum*), another montane coni-fer, were also identified in Stratum Ib. These trees presently occur only at higher elevations.

Twenty-seven plant taxa were iden-tified in the macrofossil analysis of Stratum IIa. Gambel oak leaves are the most abundant macrofossil, and this is the most common species growing outside the cave today. The leaves of barberry (*Berberis fremontii*), single leaf ash, and florets of sedge (at least three species) are common. Macrofossils of spruce, Douglas fir, and Rocky Mountain juniper persist into Stratum IIa, al-though they are rare.

RADIOCARBON DATING

Results of the radiocarbon analysis of samples from Strata Ib and IIa are presented in Table 3. Those samples from Stratum IIa and one from Ib (UGa636) were collected by the University of Utah and are discussed more fully elsewhere (Jen-nings, *Introduction*). Three other dung samples collected at grid 14R19 were ini-tially submitted to the Laboratory of Isotope Geochemistry, University of Arizona: A single piece of ruminant dung (A-1654) incorporated into and protruding from the topmost sediments of Stratum Ib was intended to date the end of Ib depo-sition. Radiocarbon dates on dung frag-ments from the top 10 cm. (A-1653) and the bottom 10 cm. (A-1660) of Stratum Ib were intended to bracket that stratum. Using Martin's overkill theory of mega-faunal extinction as a working hypothesis, we expected the youngest date from Stra-tum Ib to fall at ca. 11,000 B.P. Al-though the youngest radiocarbon date did indeed fall within this interval (11,020± 180; A-1660), the sample was recoded as coming from the **bottom** of Stratum Ib (Table 3). The topmost dung sample (as recorded in the field notes) yielded the oldest date (13,040±440; A-1654). Since there is a dating reversal in samples from a well-consolidated stratum where we believe the chances of *in situ* con-tamination are remote, we feel that, through a clerical error, samples from the top and **bottom** of Stratum Ib were switched. Another sample of dung frag-ments from the **bottom** 10 cm. of Stratum Ib was submitted for radiocarbon analysis. Assuming that the two samples had been somehow exchanged, the age of this new sample (A-1800) was expected to fall within one standard deviation of A-1654 (13,040±440). The age of A-1800 was determined to be 12,320±160 B.P., which (at 1 standard deviation) is within 120 years of A-1654.

The radiocarbon dates from Stratum Ib allow us to make several conclusions. Stratum Ib apparently represents a span of ca. 2,000 years. The youngest dates for Ib--11,810±140 B.P. (UGa636) and 11,020±180 B.P. (A-1660)--imply a hiatus of ca. 2,000 years between the cessation of Stratum Ib deposition and the depo-sition of most of IIa. The terminal age of Stratum Ib is contemporaneous with

TABLE 3

RADIOCARBON DATES FROM STRATA Ib AND IIa

Strata	Depth (m) Below Datum	Material Dated	Age (BP)	Standard Deviation 1	Standard Deviation 2	Lab No.
IIa	*	*Yucca harrimaniae* Sandal Fragments	8875	±125	±250	SI2416
IIa	*	*Quercus gambelii* Leaves	8690	±75	±150	SI2417
Ib		Ruminant Dung	11,810	±140	±280	UGa636
Ib	97.65-97.75	Ruminant Dung Single Fragment	13,040	±440	±880	A1654
Ib	97.65-97.75	Ruminant Dung Multiple Fragments	12,070	±210	±420	A1653
Ib	98.05-98.15	Ruminant Dung Multiple Fragments	11,020	±180	±360	A1660
Ib	98.05-98.15	Ruminant Dung Multiple Fragments	12,320	±160	±320	A1800

*See Jennings, *Introduction*.

the terminal dates of dung strata of the extinct shasta ground sloth in other southwestern caves (Martin 1973, Long et al. 1974, Spaulding and Martin In press).

DISCUSSION

Plant remains in Strata Ib and IIa indicate that a community with montane affinities existed near Cowboy Cave at the close of the late Pleistocene and possibly into the early Holocene. The nearest locality where spruce, Douglas fir, Rocky Mountain juniper, and birch grow together today is the Henry Mountains 50 km. to the southwest at elevations exceeding 2,200 m.

Macrofossils of at least three species of sedge (Table 2) and pollen of birch, willow, and cottonwood indicate that the small canyon below the cave supported a perennial stream. While no macrofossils of birch were found, the pollen may be that of water birch. Today, water birch is common in the southwest along streams and in moist canyons at elevations greater than 2,000 m. It may have been growing along the stream banks with willow and cottonwood.

The pattern of vegetation in the Robbers Roost area today is a mosaic of pinyon-juniper woodland on bedrock exposures, with big sagebrush communities on the flats. It is probable that an analogous pattern existed at the close of the late Wisconsin, with montane spruce and Douglas fir occupying bedrock exposures with woodland conifers. Sagebrush communities occupied the sandy flats. Wright et al. (1973) noted that the favorable habitat for big sagebrush appeared to have been greatly expanded during the late Wisconsin in the Chuska Mountains of New Mexico, 240 km. to the southeast.

The age of the initiation of the deposition of Stratum IIa is unknown, but evidence supports a hiatus of ca. 2,000 years between Stratum Ib and IIa. Despite the hiatus, the pollen samples (RPF, Fig. 1) from either side of the Ib/IIa contact are quite similar. The possibility of some mixing exists; however, Gramineae pollen, a major constituent of the dung, continues as only a trace above the contact. This, along with the compacted surface of Stratum Ib, suggests that mixing was minimal.

The top two pollen samples in Stratum IIa have less total pollen per gram than the lower two (Fig. 2). This may reflect a decrease in the productivity of the surrounding plant community, or a change in the rate of eolian deposition. Since a major phase of eolian deposition was initiated during Stratum IIa times, higher concentrations of pollen in the bottom of IIa may reflect a low initial rate of eolian deposition that increased as Holocene conditions became more pronounced. We interpret the decrease in APF of *Artemisia, Betula* and Cyperaceae pollen and the increase in *Juniperus* and *Quercus* pollen as a real change in the local vegetation.

Sedges and extralocal arboreal species persist into Stratum IIa. Although macrofossil samples represented the entire Stratum, the presence of spruce and Douglas fir macrofossils (Table 1) demonstrates that for part of the time during which Stratum IIa was deposited, these trees were growing near the cave. The pollen in IIa also indicates that the local plant community retained its mesophytic-montane character with extensive big sagebrush habitat across the Stratum Ib/IIa boundary. Juniper then expanded, followed by a dramatic change when big sagebrush decreased and oak became well-established near the cave, as it is today. Radiocarbon dates from Strata IIa indicate that this change occurred around 8700 B.P. The first cultural level above Stratum IIa yielded a radiocarbon date of 8275±80 B.P. (SI2418 on charcoal).

Martin and Mehringer (1965), in reviewing the then-existing pollen records from the southwest, placed the late Wisconsin-Holocene vegetational boundary at 12,000 B.P. They also noted (p. 434) that plant macrofossils were needed to supplement the pollen record. In later reviews, Mehringer (1967a:185-89, 1967b: 249-51), noted that conditions similar to those of the present did not occur until 7500 to 7000 B.P. The initial radiocarbon dates, pollen, and macro-fossil data from Cowboy Cave show that the local vegetation retained its mesophytic-montane character until ca. 8700 B.P. This agrees with other evidence for the persistence of mesic vegetation in the arid southwest well into the early Holocene (see summary in Van Devender 1977).

It is unlikely that during the late Wisconsin a continuous forest corridor existed between the Robbers Roost area and any area that presently supports montane vegetation (Spaulding and Van Devender 1977). Most of the land to the east along the Green River and to the west along the Dirty Devil River lies below 1,200 m. The San Rafael Desert to the north was probably an effective barrier to montane species, as was the confluence of the Colorado and Dirty Devil **Rivers** to the south.

The Cowboy Cave record suggests that montane communities were more widespread on the Colorado Plateau than they are today. Elevations above 1,700 m. on sandstone and 2,000 m. on limestone (Van Devender **et al. In press) would have** afforded favorable habitat for montane species during Pleistocene glacial stades. Long distance transport of propagules would have been effective in establishing populations and in maintaining gene flow between populations of montane species during glacial intervals.

Pollen in uppermost Strata Ia and Ib is derived from herbivore dung, with smaller amounts contributed by the local pollen rain. Stratum Ib pollen spectrum may reflect seasonality in the occupation of the cave by herbivores (cf. *Bison*). In bighorn sheep dung, high RPF of Gramineae are indicative of late summer, fall or winter diets [Spaulding 1974]. Bartos [1972] found that the summer feces of mule deer were characterized by high absolute amounts of pollen. The consistent presence of plantain (*Plantago*) pollen in the Cowboy Cave dung and the high amounts of total pollen (up to 290,000 grains/gram) argue for ingestion during the growing season.

Perhaps the bison sought shade in the large cave. Lower elevations along the Colorado River to the south would have offered better wintering grounds. The Robbers Roost area may have been part of a seasonal migratory circuit such as those of bighorn sheep (Hansen 1965, Geist 1971), elk (Capp 1968), or historic bison and caribou.

The overkill theory has been proposed by Martin (1967, 1973) to account for the extinction of Pleistocene megafauna in the western hemisphere. The abrupt disappearance of so many taxa without accompanying niche replacement is unique in the paleontological record and cannot be accounted for by natural phenomena, such as climatic change. One of the primary features of the overkill model (Mosimann and Martin 1975) is that the phenomenon should have occurred everywhere in North America at about the same time, ca. 11,000 B.P. At any one locality Pleistocene fauna and big game hunters (? Clovis culture) would have coexisted for only about ten years. Viewed as an ecological phenomenon, overkill calls for all large mammalian taxa to be affected, but not necessarily eradicated. The data from Stratum Ib appears to support Martin's overkill theory and document the effect of this phenomenon on a taxon (*Bison*) that ultimately survived Pleistocene extinction.

SUMMARY

Plant macrofossils and pollen from Cowboy Cave demonstrate the presence of montane trees such as spruce, Douglas fir, and birch and aquatics at a site which is today characterized by pinyon-juniper woodland and sagebrush flats with no permanent water. Pollen evidence indicates that big sagebrush communities may have been more extensive at the same time as montane species grew near the site. We interpret an 8700 B.P. date on oak leaves as marking the inception of a more xeric climate and as dating the approximate termination of mesophytic-montane vegetation near the Cowboy Cave site. Pollen within herbivore (cf. *Bison*) dung from Cowboy Cave indicates that the site may have been frequented during the summer. It supports Hansen's cuticle studies that disclose a grazing diet, predominantly of grasses and sedges. No similar deposits of late glacial-age, large grazing herbivore dung are known. The youngest radiocarbon date on Stratum Ib ruminant dung is contemporaneous with the youngest dates on sloth dung strata from other caves in the Southwest and supports the theory of Pleistocene over-kill proposed by Martin.

[*Acknowledgments.* Research was done at the Laboratory of Paleoenvironmental Studies, Department of Geosciences, University of Arizona, with funds provided by P. S. Martin under NSF Grant DEB 75-13944. We thank J. D. Jennings, Department of Anthropology, University of Utah, for his aid and patience. Radiocarbon dating was done with the help of A. Long and L. D. Arnold, Laboratory of Isotope Geochemistry, Department of Geosciences, University of Arizona. The assistance of T. R. Van Devender and C. V. Haynes, University of Arizona, Beverly Albee, University of Utah, and Ranger James Walters, National Park Service, was invaluable. Betty M. Fink aided in editing. This is Contribution No. 793, Department of Geosciences, University of Arizona, Tucson.]

REFERENCES

BARTOS, FRANCIS M.

1973   Pollen in Fecal Pellets As An Environmental Indicator. M. S. thesis, University of Arizona, Tucson.

BLISS, WESLEY L.

1940   A Chronological Problem Presented by Sandia Cave, New Mexico. *American Antiquity*, Vol. 5, No. 3, pp. 200-01. Washington, D. C.

1941   Reply to the Editor. *American Antiquity*, Vol. 6, No. 1, pp. 77-78, Washington, D. C.

CAPP, JOHN C.

1968   Bighorn Sheep, Elk, Mule Deer Relations. *Contributions of the Rocky Mountain Nature Associates.* Estes Park, Colorado, and Department of Fishery and Wildlife Biology, Fort Collins.

DAVIS, MARGARET B.

1963   On the Theory of Pollen Analysis. *American Journal of Science*, Vol. 261, No. 10, pp. 897-912. New Haven.

DAVIS, MARGARET B., ET AL.

1974   Calibrating Absolute Pollen Influx. In *Quaternary Plant Ecology*, H. J. B. Birks and R. G. West (Eds.), pp. 9-26. Wiley Interscience, New York.

DENTON, GEORGE H.

1974   Quaternary Glaciations of the White River Valley, Alaska. *Geological Society of America Bulletin*, Vol. 85, No. 6, pp. 871-92. Boulder.

GEIST, VALERIUS

1971   *Mountain Sheep: A Study in Behavior and Evolution.* University of Chicago Press, Chicago.

HANSEN, CHARLES G.

1965   Summary of Distinctive Bighorn Sheep Observed on the Desert Game Range, Nevada. *Desert Bighorn Council Transactions*, Vol. 9, pp. 6-10. Las Vegas.

HARRINGTON, MARK R.

1933   Gypsum Cave, Nevada. *Southwest Museum Papers* No. 8. Los Angeles.

KEARNEY, THOMAS H. AND ROBERT H. PEEBLES

1964   *Arizona Flora.* University of California Press, Berkeley.

LONG, AUSTIN, ET AL.

1974   Extinction of the Shasta Ground

Sloth. *Geological Society of America Bulletin*, Vol. 85, No. 12, pp. 1843-48. Boulder.

MAHER, LOUIS J., JR.

1972　Nomograms for Computing 0.95 Confidence Limits of Pollen Data. *Review of Paleobotany and Palynology*, Vol. 13, No. 2, pp. 85-93.

MARTIN, PAUL S.

1963　*The Last 10,000 Years: A Fossil Pollen Record of the American Southwest.* University of Arizona Press, Tucson.

1967　Prehistoric Overkill. In *Pleistocene Extinctions*, Paul S. Martin and Herbert E. Wright, Jr. (Eds.), pp. 75-120. Yale University Press, New Haven.

1973　The Discovery of America. *Science*, Vol. 179, No. 4077, pp. 969-74.

MARTIN, PAUL S., ET AL.

1961　Rampart Cave Coprolite and Ecology of the Shasta Ground Sloth. *American Journal of Science*, Vol. 259, No. 2, pp. 102-27. New Haven.

MARTIN, PAUL S., AND PETER J. MEHRINGER, JR.

1965　Pleistocene Pollen Analysis and Biogeography of the Southwest. In *The Quaternary of the United States*, Herbert E. Wright, Jr. and David G. Frey (Eds.), pp. 433-51. Princeton University Press, Princeton.

MEHRINGER, PETER J., JR.

1967a　Pollen Analysis of the Tule Springs Area, Nevada. *In* "Pleistocene Studies in Southern Nevada", H. Marie Wormington and Dorothy Ellis (Eds.), *Nevada State Museum Anthropo-*

logical Papers No. 13, pp. 129-200. Reno.

1967b　The Environment of Extinction of the Late Pleistocene Megafauna in the Arid Southwestern United States. In *Pleistocene Extinctions*, Paul S. Martin and Herbert E. Wright, Jr. (Eds.), pp. 247-66. Yale University Press, New Haven.

MOSIMANN, JAMES E.

1965　Statistical Methods for the Pollen Analyst: Multinomial and Negative Multinomial Techniques. In *Handbook of Paleontological Techniques*, Bruno Kunnel and David Raup (Eds.), pp. 636-73. W. H. Freeman and Co., New York.

MOSIMANN, JAMES E., AND PAUL S. MARTIN

1975　Simulating Overkill by Paleo-Indians. *American Scientist*, Vol. 63, No. 3, pp. 304-13. New Haven.

SANGSTER, A. G., AND H. M. DALE

1961　A Preliminary Study of Differential Pollen Grain Preservation. *Canadian Journal of Botany*, Vol. 39, No. 1, pp. 35-43. Ottawa.

SHEPPARD, JOHN C.

1975　A Radiocarbon Dating Primer. *Washington State University College of Engineering Bulletin* No. 338. Pullman.

SPAULDING, W. GEOFFREY

1974　*Pollen Analysis of Fossil Dung of Ovis canadensis from Southern Nevada.* M. S. thesis, University of Arizona, Tucson.

SPAULDING, W. GEOFFREY, AND PAUL S. MARTIN

Ground Sloth Dung of the Guadalupe Mountains. In *Biological Investigations in the Guadalupe*

Mountains National Park, Texas, Symposium Volume, Hugh H. Genoways and Robert L. Baker (Eds.), U. S. National Park Service, Washington, D. C. (In press)

SPAULDING, W. GEOFFREY AND T. R. VAN DEVENDER

1977   Late Pleistocene Montane Conifers in Southeastern Utah. *Southwestern Naturalist,* Vol. 22, No. 2, pp. 269-71.

STOCKMARR, JENS

1971   Tablets With Spores Used in Absolute Pollen Analysis. *Pollen et Spores,* Vol. 13, No. 4, pp. 615-21. Montpellier.

1973   Determination of Spore Concentration With An Electronic Particle Counter. *Geological Survey of Denmark, 1972 Yearbook,* pp. 87-9. Copenhagen.

VAN DEVENDER, THOMAS R.

1977   Holocene Woodlands in the Southwestern Deserts. *Science,* Vol. 198, No. 4313, pp. 189-92. Washington, D. C.

VAN DEVENDER, THOMAS R., ET AL.

n.d.   Late Pleistocene Plant Communities in the Guadalupe Mountains, Culberson County, Texas. In *Biological Investigations in the Guadalupe Mountains National Park, Texas, Symposium Volume,* Hugh H. Genoways and Robert L. Baker (Eds.). U. S. Park Service, Washington, D. C. (In press)

WRIGHT, HERBERT E., JR., ET AL.

1973   Present and Past Vegetation of the Chuska Mountains, Northwestern New Mexico. *Geological Society of America Bulletin,* Vol. 84, No. 4, pp. 1155-80. Boulder.

APPENDIX III
LATE PLEISTOCENE PLANT FRAGMENTS IN THE DUNGS OF HERBIVORES AT COWBOY CAVE

Richard M. Hansen

For half a century, preserved organic fragments of plants and animals have been providing records of late Pleistocene biotic distributions. The sudden extinction of certain large herbivores closely following the chronology of prehistoric man's spread during times of climatic changes confounds the proof for the cause of extinctions. The dry dungs and bones of extinct mammals beneath the cultural layer in Cowboy Cave could provide undiscovered evidence useful for interpreting the late Pleistocene events. A need exists for clear evidence of human activity associated with the remains of animals which might have been exterminated by Paleo-Indians as they became established in new areas.

The objectives of this study were to identify the fragments of plants found in various dung samples, to identify the animals which made the defecations, and to report these findings as accurately as possible.

MATERIALS

Cowboy Cave is in the Navajo Sandstone cliffs facing southeast along a short, nameless, box canyon near Canyonlands National Park. The cave is in an area administered by the Bureau of Land Management. The dung samples supplied to me were obtained during excavations conducted by the Department of Anthro-

pology, University of Utah, June 3 through July 26, 1975 [Jennings 1975]. In addition, two lists of plant species now present in the environs of Cowboy Cave were provided by Jennings, and a list with some additions was provided by Spaulding and Van Devender (see *Appendix I*). Spaulding also provided a list of macrofossil plants he had tentatively identified, and Petersen (see *Appendix II*) provided a table of fossil pollen counts from Stratum IIa and the herbivore dung layer, Stratum Ib [Jennings 1975, Fig. 5].

METHODS

Plants found in fragmentary condition were identified and classified in a manner similar to the procedures used by Laudermilk and Munz (1934, 1938) in studies of Shasta ground sloth dung from Gypsum, Rampart, and Muav caves. Previous experience with fossil dungs from caves in the southwestern United States suggested that the procedures in handling modern herbivore fecal samples (Sparks and Malechek 1968; Ward 1970; Storr 1961; Williams 1969) were less satisfactory for the dry, brittle fragments in the Cowboy Cave samples. The samples for micro-histological analyses were presoaked in equal volumes of ethyl alcohol, water and glycerine for several days to soften the epidermal tissues so they became elastic. The soaked material was washed over a screen with 1 mm. openings and the smaller epidermal fragments were collected over a screen having 0.1 mm. openings. The large fragments were placed in a blender with water and agitated to dislodge epidermal fragments from any stems and to reduce the size of leaf fragments to 1 mm. or less in size. Alternate blendings and washings were used to obtain the necessary epidermal materials for making microscope slides according to the procedures reported by Sparks and Malechek (1968), Ward (1970) and Flinders and Hansen (1972).

Plants of different species may fragment differently when chewed and digested, and grinding fecal samples or blending them renders uniformity of fragment sizes between species of plants and between species of herbivore. The resulting smaller fragments permit thin microscope slides to be made which transmit more light and make identification easier and more accurate. Fragments of plants of similar size (less than 1 mm.) represent their original dry weight in the ingesta quite satisfactorily (Dearden et al. 1975). The 1:1 ratio in size for those items that fragment differently permits quantifications from microscope slides to be done most efficiently (Sparks and Malechek 1968).

Five microscope slides were made for each fecal sample which was to be examined and 20 slides were made from the bulk sample. There were 20 fields examined at 100X per slide. The relative density percentage of classified plant fragments appears to be a good approximation of the relative dry weight of each food category in the diets of herbivores (Hansen et al. 1973; Todd and Hansen 1973). Anthony and Smith (1974) found that quantifications of deer diets by microscopic analysis of feces gave very similar results to those of rumen samples. Dearden et al. (1975) reported that discerned fragments in digested and simulated digested residues were not greatly different for reindeer, cattle, and bison.

The reference collection of the Composition Analysis Laboratory for epidermis of plants was used for classifying fragments found in the samples. The reference slides available contained most of the genera but was deficient for some species in the modern vegetation of the area. Many kinds of plants from various vegetation zones of North America were available on microscope slides in the reference collection. Fragments were classified in samples according to how well they matched fragments on the reference slides.

The seeds from the samples were identified by the technicians of the Colorado State Seed Laboratory by comparing the fossil seeds with vouchered reference seed samples in their collection.

Hairs taken from the bulk sample were classified if their characteristics matched those in reference collections in the Wyoming Game and Fish Laboratory and the Composition Analysis Laboratory. When reference material was lacking, the descriptions of hair characteristics and photographs in Moore et al. (1974) were used. The procedures described by Moore were followed for studying hair casts, scale patterns, and hair characteristics by microscopic examinations.

In addition to the samples prepared for microhistological analysis, some Cowboy Cave dung samples were processed to determine relative dry weights of plant fragments. Each sample was individually washed through four nested screens with openings sized 1,000, 500, 250, and 100 microns. Fractions remaining on each screen after complete, but gentle, washing were placed in a drying oven at 65° C until dry, and then weighed to a 1 mg. accuracy.

The dry weight proportions of different-sized fragments in 67 different dung samples of herbivores was determined. The body sizes of the herbivores ranged from *Elephas maximus* to *Neotoma mexicana* and included a variety of monogastrics and ruminants. The size/dry weight relationships of fragments in the dungs were considered as one criterion for dung identifications of Cowboy Cave samples. A reference collection of fecal material was available to make comparisons for the dung fecal pellet samples from Cowboy

Cave (Fig. 1).

BESTIARY OF CAVE

Direct evidence for the identification of animals in the compacted stratum of pink sand and finely chopped vegetal matter came from skeletal remains, hair, and dung. When all three elements are present for a particular animal the likelihood of its presence is reinforced. The number of times these elements are encountered may be used to suggest the relative importance of the animals found, provided the collections and examinations were done without bias.

Bison dung samples obtained from Stratum Ib frequently had holes from coprophagous larval insects. The botanical composition of the fecal samples classified as bison consisted primarily of grasses and sedges (Table 1), not much different from modern bison now living at Colorado National Monument, Colorado. An estimated 90 percent of the hairs recovered from the bulk sample appeared to be bison, though not all of them received a detailed microscopic examination. Some hairs were underfur and others were typical guard hairs. Jennings [1975] reported finding bones from extinct bison in Cowboy Cave.

The fractional dry weights of plant fragments from bison dung were similar to modern bison feeding in late spring or early summer. The dry weight of fragments > 1,000 microns averaged 14 percent, which is below average (Table 2). Plants in an early growth stage are more easily fragmented during chewing and digestion than are mature plants. The large fragment category was less than average in percentage dry weight, which suggests the food was in an early growth stage when ingested.

Samples of mammoth dung were obtained from Stratum Ib. One piece of mammoth dung had a 2 to 3 mm. black piece of material appearing to be charcoal imbedded in its surface. An elephant tusk 12 cm. in length was recovered from the base of Stratum Ib [Jennings 1975]. Three dung samples from the stratum were separated by fragment sizes and the fraction > 1,000 microns averaged 62 percent of the dry weight, which is similar to that of a modern elephant (Table 2). The botanical composition of the dung samples was ≃99 percent Gramineae fragments (Table 1). One spine resembling those of *Crateagus* sp. was found in the mammoth dung. Under natural conditions elephants eat primarily grasses and of 108 droppings from the African elephant (*Loxodonta africana*) grasses averaged 93 percent (Wing and Buss 1970). If Cowboy Cave had been used much by a mammoth, little time would have been needed to fill the cave; the African elephant eliminates about 136 kg. of dung per day (Wing and Buss 1970).

A few loose pieces of extinct horse dung of similar texture were also from Stratum Ib. Portions of two large pieces were studied for botanical composition and fragment sizes. Some of the samples have larvae holes from dung-feeding insects. The external morphology needed for more positive identification was destroyed by trampling and compaction.

Firm evidence of horses associated with the stratum, however, is lacking. No skeletal remains were found, but certain unclassifiable hair fragments could be from horses. Some of the dung samples have the appearance and texture of compacted horse dung. The two samples examined for botanical composition contained 99 percent Gramineae fragments (Table 1) which are the kinds of plants preferred by wild horses (Hubbard and Hansen 1976). The fraction of the samples > 1,000 microns averaged 50 percent of the dry weights, which is in the probable range for wild horses and burros (Table 2). The major grass in the horse dung was the same as that in the mammoth dung. Without additional evidence, it is possible that the "horse" and "mammoth" dungs in this study came from the same mammal herbivore.

Two pieces of ground sloth dung of identical appearance came from Stratum Ib. Each had holes from larvae of coprophagous insects. Each appeared to have a poorly defined mucus coating. Externally, the pieces resembled sloth or

FIG. 1. a, Artiodactyl type pellets (two lower rows) recovered from late Pleistocene deposits of Cowboy Cave, Utah, and the pellets from living bighorn sheep
(S2), mule deer (D3), and elk (E2); b, three very large (elk-camel) pellets
from the late Pleistocene deposits of Cowboy Cave; c, the apparent dungs of
extinct elephant (left) and extinct horse (right) from the Cowboy Cave deposits;
d, twigs apparently from the dungs of Shasta ground sloth and extinct elephant
from Cowboy Cave (recovered from a bulk sample); e, one-half of a segment of
apparent Shasta ground sloth dung (left) recovered from Cowboy Cave and a segment
of Shasta ground sloth dung (right) from Rampart Cave, Arizona.

TABLE 1

BOTANICAL COMPOSITION OF LATE PLEISTOCENE HERBIVORE
DUNGS OF COWBOY CAVE, WAYNE COUNTY, UTAH

Plant Names*	Average Percent in Dungs			
	Bison (N=4)	Mammoth (N=3)	Horse (N=2)	Sloth (N=2)
Agropyron	1	1		
Carex	12	<1	<1	
Festuca		<1		
Hilaria		<1		
Oryzopsis	4	1		<1
Sporobolus	73	95	98	1
Sporobolus Seed	1	<1	<1	
Stipa	5	2	1	
Amelanchier Type				36
Chenopodium Type	<1	<1		<1
Clematis Type	<1			
Equisetum	2	1		
Pseudotsuga				60
Sphaeralcea		<1	<1	
Legume Stem	<1			
Unidentified Forb	1			
Forb Seed				3

*Tentative names based on types of tissue observed.

TABLE 2

THE RELATIVE PERCENTAGE DRY WEIGHTS OF FOUR SIZES OF FRAGMENTS FOUND IN THE FECES OF SEVERAL KINDS OF HERBIVORES ($\bar{X} \pm$ SD)

Names of Herbivores	Screen Size in Microns				N
	>1000	>500 <1000	>250 <500	>100 <250	
Nothrotheriops shastense	65 ± 13	8 ± 5	16 ± 6	11 ± 5	6
Elephas maximus	62	11	15	12	2
Equus asinus	59 ± 7	18 ± 4	12 ± 2	11 ± 3	6
Equus caballus	50 ± 3	20 ± 2	18 ± 1	12 ± 4	3
Gopherus agassizii	42	15	15	28	2
Ovibos moschatus	28	20	24	28	1
Sauromalus obesus	27	32	21	20	1
Rangifer tarandus sylvestris	21 ± 2	27 ± 1	26 ± 1	26 ± 1	3
Box taurus	20	19	27	34	2
Bison bison	19 ± 5	18 ± 4	27 ± 4	36 ± 8	7
Cervus canadensis	19 ± 5	22 ± 9	30 ± 7	29 ± 7	4
Erethizon dorsatum	17	29	33	21	1
Alces americana	15	33	30	22	2
Cynomys ludovicianus	14	27	36	23	1
Lama glama	13	26	28	33	1
Lepus townsendii	11	31	23	35	1
Odocoileus hemionus	11 ± 4	21 ± 6	35 ± 3	33 ± 5	3
Ovis aires	10	22	34	34	2
Oreamnos americanus	9 ± 9	23 ± 9	31 ± 3	37 ± 12	5
Ovis canadensis nelsoni	8 ± 5	27 ± 1	35 ± 3	30 ± 3	4
Sylvilagus auduboni	7 ± 4	27 ± 3	38 ± 6	28 ± 3	3
Lepus californicus	6	30	31	33	1
Antilocapra americana	6	25	34	35	2
Ovis canadensis canadensis	5	24	38	33	2
Neotoma mexicana	5 ± 3	29 ± 8	33 ± 2	33 ± 8	2

horse dung. One piece was studied and the other was saved for reference.

Five hair fragments which strongly resembled Shasta ground sloth hairs in width, color, and scale pattern were found in the bulk sample. Sloth hairs resemble bovid hairs and in the hair parts examined no distinguishing oval bodies were present. The sloth dung examined was similar to the Shasta ground sloth for the > 1,000 micron fractional percentage (Table 2). The fraction was 60 percent in this study. The plant fragments in the sloth dung were primarily from Douglas fir needles and serviceberry leaves (Table 1). The size (5 by 10 by 3 cm.) of the specimen, the dry weight of the > 1,000 micron fraction, and the botanical composition of the sample strongly rule out ruminants and equines as the animal which made the defecations. Spaulding and Martin (1976) reported Douglas fir needles in the dung of the Shasta ground sloth from Cave 08 of the Guadalupe Mountains, Texas.

Fecal pellets of three kinds of browsing herbivores were recovered. Six pellets averaged (±SD) 1.35±0.20 gm. dry weight and are much larger than modern elk pellets. I have tentatively classified the six pellets as "camel-elk;" they are smaller than those of a dromedary camel and not shaped like those of a moose (*Alces*). Their dry weight fraction > 1,000 microns averages 11 percent, which is similar to the cervids, moose, and llama (Table 2). The majority of the plant fragments were serviceberry, which further suggests they were not from elk, which eat primarily grasses (Hansen and Clark 1977).

Twenty-nine fecal pellets resembling those of elk or deer were obtained from Stratum Ib. The pellets were essentially similar and averaged (±SD) 0.47± 0.05 gm. in dry weight. Some of the pellets were saved and the remainder were studied for botanical composition (Table 3) and fragment sizes. The plant fragments in these pellets were primarily serviceberry, which is an important spring and summer food of mule deer (Hubbard and Hansen 1976).

Fecal pellets resembling deer or bighorn sheep were also found. One hair from a bighorn was identified from the bulk sample, and skeletal remains of modern bighorns were found in the uppermost layers of the cultural deposits [Jennings 1975]. One of the pellets was used in making microscope slides, and five similar pellets were also made into microscope slides. They averaged 0.27 gm. each in dry weight. The plant fragment components were primarily serviceberry (Table 3). Since bighorn sheep show a preference for grasses, the fecal pellets were probably from mule deer.

Three small, consolidated pieces of ruminant dung contained plants of the same kinds found in the camel-cervid-bovid types (Table 3) and their dry weight fraction > 1,000 microns averaged only 4 percent. Herbivores feeding on lush vegetation in the early summer frequently do not form pellets, and so the ruminant-like dung samples probably came from an herbivore about the size of an elk or deer. Because all of the camel-cervid-bovid type pellets contained a preponderance of browse and forbs, it is unlikely that these dungs came from elk or bighorns, which feed primarily on grasses and sedges (Hansen and Clark 1977; Hansen and Reid 1975).

One pellet from Stratum Ib was clearly identified as cottontail (*Sylvilagus* sp.). Three cottontail hairs were also identified (FS-630). The majority of the fragments in the pellet could not be classified because reference materials were not available. However, the identified fragments were similar to those reported in the diets of cottontails from elsewhere (Hansen and Gold 1977; Turkowski 1975).

About five liters of "bulk sample" material was taken from Stratum Ib. The sample was primarily pink sand containing small fragments of plants assumed to have originated from herbivore dung. A few pieces of dung were up to 2 cm. in dia. The organic materials were separated from the sand by sifting through screens and by water flotation. The bulk sample most closely resembled bison dung in plant species composition. The remainder of the organic material was searched for

TABLE 3

BOTANICAL COMPOSITION OF LATE PLEISTOCENE HERBIVORE DUNGS OF COWBOY CAVE, WAYNE COUNTY, UTAH

Plant Names*	Average Percent in Dungs				
	Camel-elk** (N=3)	Elk-deer** (N=8)	Deer-bighorn** (N=2)	Ruminant (N=3)	Cottontail (N=1)
Agropyron	<1	<1		<1	
Carex		1			
Festuca	<1				
Sporobolus	1	4	<1	2	
Sporobolus Seed		<1			
Stipa		<1		<1	
Artemisia					<1
Amelanchier Type	76	87	98	97	
Chenopodium Type	2	8	2		3
Chenopodium Seed		<1	<1		
Composite	8	<1			
Cryptantha Type		<1			
Equisetum	<1				6
Erogonum					10
Physaria		<1			
Quercus		<1			
Sphaeralcea		<1			1
Forb Seed	12	1			1
Seed Heads					78
Moss		<1			
Legume Stem		<1			
Pod		1			

*Tentative names based on types of tissue observed.

**Composites of several pellets.

mammal hairs, seeds, and macrofossils using low power magnifications.

The seeds in the bulk sample were: *Sporobolus cryptandrus, Oryzopsis hymenoides, Carex* sp., *Cyperus* sp., *Malva* sp., *Rosa* sp., *Chenopodium berlandieri, Tradescantia* sp., *Franseria* sp., *Picea* sp., *Corispermum* sp., *Amelanchier* sp., and some unclassified. Large plant fragments were: *Picea* sp. (leaves), *Artemisia tridentata* (wood), *Pseudotsuga menziesii* (leaves), and *Opuntia* sp. (areoles).

Canidae hairs were identified but the specimens were not so distinct that they could be classified to genus. Six fragments of hairs resembled human hairs. They were lightly pigmented, very thin, and the medula was absent. The scales were irregular, waved mosaic with margins smooth to crenate. The six hair fragments were encrusted with mineral and organic deposits, and they were brittle. These hairs lacked the strong elastic quality characteristic of fresh hairs and were not considered to be recent contaminants. The unidentified hairs in our collection may be tentatively classified as unknown, bisonlike and humanlike. With more study using additional reference materials some of the unknown hair fragments may yet be identified.

DISCUSSION

There is strong evidence that the samples examined in this study originated from an environment considerably different from the present one at the Cowboy Cave site. The fecal pellets of browsers were generally more amorphous than are fall or winter pellets, implying that the range was being used primarily in the summer. The holes of maggots in the dungs of the larger herbivores indicates the area was being used at the warmest period of the year. The leaves of deciduous shrubs and forbs and the predominance of flower parts in herbivore diets also suggests summer use. The megafauna evidence suggests this area contained a mixture of plant species typical of open parklands in the lower spruce-fir zone. The lack of juniper, pine, and sagebrush in pellets of cervidlike herbivores suggests the area was cold and had deep snow in the winter. The fact that the organic materials buried in the sands were not completely decomposed suggests that the area had little rain during what may have been hot, dry summers.

Because of the scarcity of data, some controversial hypotheses have been proposed to describe when the earliest Paleo-Indians entered the New World (MacNeish 1976) and to address the problem of Pleistocene megafauna extinctions (Martin 1973; Mosimann and Martin 1975). There is still a lack of good evidence, however, to associate Paleo-Indian artifacts and extinct large mammals before about 10,000 years ago. At about this time, a series of relatively specialized tools for hunting and butchering large animals can be found in New World deposits.

[*Acknowledgments*. It is with some pleasure that I express appreciation to many persons who were helpful in this research. Jesse D. Jennings of the University of Utah provided the samples and the opportunity for me to become involved. W. G. Spaulding and T. R. Van Devender, Geosciences Department, University of Arizona, and K. L. Petersen, Department of Anthropology, Washington State University, provided plant lists and results of pollen and dung analysis. Tommy Moore, Wyoming Game and Fish Laboratory, University of Wyoming, gave valuable advice and identified hairs. A. L. Larsen, Colorado State Seed Laboratory, Colorado State University, made seed identification and confirmations. Members of the Composition Analysis Laboratory, Colorado State University, were essential and their efforts are appreciatively acknowledged: Theresa M. Foppe identified the plants and seeds, Mark K. Johnson classified the hairs, Hal K. Reynolds separated various dungs for fragment sizes. The Graduate School, Colorado State University provided some funds for this study (Faculty Grant No. 0338).]

REFERENCES

ANTHONY, ROBERT G., AND NORMAN S. SMITH

1974   Comparison of Rumen and Fecal Analysis to Describe Deer Diets. *Journal of Wildlife Management*, Vol. 38, No. 3, pp. 535-40. Washington, D. C.

DEARDEN, BOYD L., ROBERT E. PEGAU, AND RICHARD M. HANSEN

1975   Precision of Microhistological Estimates of Ruminant Food Habits. *Journal of Wildlife Management*, Vol. 39, No. 2, pp. 402-07. Washington, D. C.

FLINDERS, JERRAN T., AND RICHARD M. HANSEN

1972   Diets and Habitats of Jackrabbits in Northeastern Colorado. *Range Science Department Series, Colorado State University*, No. 12, p. 29. Fort Collins.

HANSEN, RICHARD M., AND RICHARD C. CLARK

1977   Foods of Elk and Other Ungulates at Low Elevations in Northwestern Colorado. *Journal of Wildlife Management*, Vol. 41, No. 1, pp. 76-80. Washington, D. C.

HANSEN, RICHARD M., AND ILYSE K. GOLD

1977   Blacktail Prairie Dogs, Desert Cottontails, and Cattle Trophic Relations on Shortgrass Range. *Journal of Range Management*, Vol. 30, No. 3, pp. 210-14. Denver.

HANSEN, RICHARD M., DONALD G. PEDEN, AND RICHARD W. RICE

1973   Discerned Fragments in Feces Indicates Diet Overlap. *Journal of Range Management*, Vol. 26, No. 2, pp. 103-05. Denver.

HANSEN, RICHARD M., AND LORIN D. REID

1975   Diet Overlap of Deer, Elk, and Cattle in Southern Colorado. *Journal of Range Management*, Vol. 28, No. 1, pp. 43-7. Denver.

HUBBARD, RICHARD E., AND RICHARD M. HANSEN

1976   Diets of Wild Horses, Cattle, and Mule Deer in the Piceance Basin, Colorado. *Journal of Range Management*, Vol. 29, No. 5, pp. 389-92. Denver.

JENNINGS, JESSE D.

1975   **Preliminary Report: Excavation of Cowboy Caves.** Report on file, Department of Anthropology, University of Utah. Salt Lake City.

LAUDERMILK, J. D., AND PHILLIP A. MUNZ

1934   Plants in the Dung of *Nothrotherium* from Gypsum Cave, Nevada. *Carnegie Institution of Washington Publication*, No. 453, pp. 29-37. Washington, D. C.

1938   Plants in the Dung of *Nothrotherium* from Rampart and Muav Caves, Arizona. *Carnegie Institution of Washington Publication*, No. 487, pp. 271-81. Washington, D. C.

MACNEISH, RICHARD S.

1976   Early Man in the New World. *American Scientist*, Vol. 64, No. 3, pp. 316-27. New Haven.

MARTIN, PAUL S.

1973   The Discovery of America. *Science*, Vol. 179, No. 4077, pp. 969-74. Washington, D. C.

MOORE, TOMMY D., LITER E. SPENCE, AND CHARLES E. DUGNOLLE

1974   Identification of the Dorsal Guard Hairs of Some Mammals of Wyoming. *Wyoming Game and Fish Department Bulletin* No. 14. Cheyenne.

MOSIMANN, JAMES E., AND PAUL S. MARTIN

  1975   Simulating Overkill by Paleo-
          Indians. *American Scientist*,
          Vol. 63, No. 3, pp. 304-13.
          New Haven.

SPARKS, DONNIE R., AND JOHN C. MALECHEK

  1968   Estimating Percentage Dry
          Weight in Diets Using a Micro-
          scope Technique. *Journal of
          Range Management*, Vol. 21, No.
          4, pp. 264-65. Denver.

SPAULDING, W. GEOFFREY, AND PAUL S.
MARTIN

  1976   Ground Sloth Dung of the Guada-
          lupe Mountains. **In** *Biological
          Investigations in the Guadalupe
          Mountains National Park, Texas,
          Symposium Volume*. Hugh H. Geno-
          ways and Robert L. Baker (Eds.).
          U. S. National Park Service
          Publication. (In press)

STORR, G. M.

  1961   Microscopic Analysis of Faeces,
          a Technique for Ascertaining
          the Diet of Herbivorous Mammals.
          *Australian Journal of Biologi-
          cal Science*, Vol. 14, No. 1,
          pp. 157-64. Melbourne.

TODD, JEFFERY W., AND RICHARD M. HANSEN

  1973   Plant Fragments in the Feces of
          Bighorns as Indicators of Food
          Habits. *Journal of Wildlife
          Management*, Vol. 38, No. 3, pp.
          363-66. Washington, D. C.

TURKOWSKI, FRANK J.

  1975   Dietary Adaptability of the
          Desert Cottontail. *Journal of
          Wildlife Managment*, Vol. 39,
          No. 4, pp. 748-56. Washington,
          D. C.

WARD, ANGUS L.

  1970   Stomach Content and Fecal
          Analysis: Methods of Forage
          Identification. *U.S.D.A. Mis-
          cellaneous Publications*, No.
          1147, pp. 146-58. Washington,
          D. C.

WILLIAMS, OWEN B.

  1969   An Improved Technique for Iden-
          tification of Plant Fragments
          in Herbivore Feces. *Journal of
          Range Management*, Vol. 22, No.
          1, pp. 51-2. Denver.

WING, LARRY D., AND IRVEN O. BUSS

  1970   Elephants and Forests. *Wildlife
          Monographs*, No. 19, pp. 1-92.
          Washington, D. C.

APPENDIX IV
CORN FROM COWBOY CAVE

Joseph C. Winter

A small sample of kernels (470) from the maize collection from Cowboy Cave was examined. The kernels, recovered from a cache of two hide bags (see *Animal Skin Bags*, Fig. 16), are very uniform in color, size, and shape, which suggests that the sample represents seed corn saved from a single harvest. Most of the kernels have a thick, flinty endosperm, but some pops and flours are also present. They are not dented or pointed like Fremont Dent, nor are they similar to the early historic types grown by the Hopi and other Southwestern groups.

The kernels are relatively small (see accompanying table) and, in size at least, unlike the larger kernels grown in late prehistoric times. On the basis of measurement alone, they are similar to the kernels of the lowest levels of Clydes Cavern (Winter and Wylie 1974) that are Archaic in context.

Corn has been found in a number of Archaic sites in the Southwest dating as early as 3000 B.P. Corn pollen is reported at En Medio Rockshelter in New Mexico (Irwin-Williams and Tompkins 1968), Cienega Creek and Tularosa Cave in Arizona (Martin, Rinaldo **et al.** 1952; Martin and Schoenwetter 1960), and LoDaiska in Colorado (Irwin and Irwin 1959). The occurrence of maize in a ca. 1600 B.P. context (*Introduction*, Table 3) at Cowboy Cave is interesting because it is so far north on the Colorado Plateau. Corn of comparable age (dendrochronological dates of A.D. 49-380) in the general area was found at the Durango Basketmaker II sites, and at Basketmaker II sites to the west on Cedar Mesa. The corn from Unit V at Cowboy Cave in a late Archaic/pre-Basketmaker context provides additional evidence that scattered groups throughout the Southwest practiced rudimentary horticulture in a wide range of environments.

COMPARATIVE KERNEL MEASUREMENTS

Site	No. of Kernels	Height	Width	Thickness
Cowboy Caves	470	6.3 mm.	7.2 mm.	4.2 mm.
Clydes Cavern				
Level 2	41	7.2	7.2	4.5
Level 3	136	7.8	7.8	4.9
Level 4	113	7.1	8.2	4.2
Levels 5-7	40	7.3	8.3	4.0

REFERENCES

IRWIN, CYNTHIA, AND HENRY IRWIN

1959   Excavations at the LoDaiska
Site. *Denver Museum of Natural
History Proceedings* No. 8.
Denver.

IRWIN-WILLIAMS, CYNTHIA, AND S. TOMPKINS

1968   Excavations at En Medio Rockshelter, New Mexico. *Eastern
New Mexico University Contributions in Anthropology*, Vol.
1, No. 2. Portales.

MARTIN, PAUL S., ET AL.

1952   Mogollon Cultural Continuity
and Change: A Stratigraphic
Analysis of Tularosa and
Cordova Caves. *Fieldiana:
Anthropology*, Vol. 40. Chicago.

MARTIN, PAUL S., AND JAMES SCHOENWETTER

1960   Arizona's Oldest Cornfield.
Science, Vol. 132, No. 3418,
pp. 33-4. Washington, D. C.

WINTER, JOSEPH C., AND HENRY G. WYLIE

1974   Paleoecology and Diet at Clydes
Cavern. *American Antiquity*, Vol.
39, No. 2, pp. 304-14. Washington, D. C.

APPENDIX V
MACROFOSSILS IDENTIFIED FROM FIRST BULK
SAMPLE OF LEAF LAYER, STRATUM IIa

Beverly J. Albee

Agropyron or *Elymus* sp.

Ambrosia acanthicarpa

Amelanchier sp.

Amelanchier utahensis

Arenaria sp.

Artemisia cf. *bigelovii*

Berberis fremontii

Brickellia microphylla

Cercocarpus intricatus

Ephedra nevadensis

Ephedra torreyana

Erigeron spp.

Fraxinus anomalas

Hackelia cf. *floribunda*

Hilaria jamesii

Juniperus osteosperma

Lomatium grayi

Oryzopsis hymenoides

Populus fremontii

Quercus gambelii

Sporobolus sp.

Stephanomeria spp.

Streptanthella longirostris

APPENDIX VI
COMPARATIVE IDENTIFICATIONS, SECOND
MACROFOSSIL SAMPLE OF LEAF LAYER,
STRATUM IIa

Beverly J. Albee

List from Table 2, *Appendix II*	Comment by Beverly Albee
Carex sp.	*Hilaria jamesii*
Carex sect. *Paludosae*	*Ephedra torreyana*
Equisetum sp.	Yes
Stipa sp.	Yes
Chenopodium sp.	Yes
Gutierrezia sp.	*Brickellia microphylla*
Yucca sp.	Yes
Corispermum sp.	Yes
Juniperus scopulorum	Yes
Agropyron sp.	Yes
Oryzopsis hymenoides	Yes
Shepherdia rotundifolia	No, hairs on *Shepherdia* star-shaped-- I couldn't determine this.
Cornus stolonifera	Yes
Unknown #8 (*Populus*?)	Yes, *Populus fremontii*.
Rosa woodsii	Yes
Pseudotsuga menziesii	Yes
Picea cf. *engelmanni*	Yes, or *Picea pungens*.
Purshia tridentata	No, probably a Composite.
Opuntia macrorhiza	Yes, not sure about species.
O. erinacea	Yes, not sure about species.
Cercocarpus intricatus	Yes
Rhus trilobata	Yes

List from Table 2 (continued)

Ostrya knowltoni

cf. *Rhamnus smithii*

Comment by Beverly Albee

Juniperus scopulorum

I would say *Rhamnus betulaefolia*, as it is the species most common in Southern Utah. Beautiful identification.

APPENDIX VII
MODERN VEGETATION INVENTORY: CANYON COMMUNITY — THE SPUR CANYON AND TRIBUTARY CANYONS INCLUDING COWBOY CAVE

Beverly J. Albee

Anacardiaceae
  Rhus radicans
  Rhus trilobata

Apocynaceae
  Apocynum cannabinum

Asclepiadaceae
  Asclepias macrosperma

Berberidaceae
  Berberis fremontii

Boraginaceae
  Cryptantha crassisepala
  Cryptantha flava
  Cryptantha jamesii
  Lappula redowskii
  Lithospermum incisum

Cactaceae
  Mammillaria tetrancistra
  Opuntia erinacea

Capparidaceae
  Cleome lutea

Caryophyllaceae
  Arenaria eastwoodiae

Compositae
  Artemisia carruthii
  Artemisia dracunculus
  Artemisia ludoviciana
  Brickellia microphylla
  Brickellia oblongifolia
  Chaenactis stevioides
  Chrysothamnus nauseosus
  Erigeron utahensis
  Gutierrezia sarothrae

Compositae (cont.)
  Haplopappus armerioides
  Haplopappus nuttallii
  Hymenopappus filifolius
  Hymenoxys acaulis
  Leucelene ericoides
  Malacothrix sonchoides
  Senecio longilobus
  Senecio multilobatus
  Solidago spp.
  Townsendia incana
  Xylorhiza tortifolia

Cornaceae
  Cornus stolonifera

Cruciferae
  Descurainia richardsonii
  Erysimum asperum
  Lepidium montanum
  Physaria australis
  Stanleya pinnata
  Streptanthella longirostris

Cupressaceae
  Juniperus osteosperma

Elaeagnaceae
  Shepherdia rotundifolia

Ephedraceae
  Ephedra nevadensis
  Ephedra torreyana

Euphorbiaceae
  Euphorbia fendleri

Fagaceae
  Quercus gambelii

Gentianaceae
 Frasera paniculata

Gramineae
 Aristida longiseta
 Festuca octoflora
 Hilaria jamesii
 Muhlenbergia pungens
 Oryzopsis hymenoides

Hydrophyllaceae
 Phacelia crenulata
 Phacelia ivesiana

Leguminosae
 Astragalus ceramicus
 Astragalus desperatus
 Astragalus lentiginosus
 Astragalus mollissimus
 Glycyrrhiza lepidota
 Lupinus pusillus
 Petalostemon flavescens
 Psoralea lanceolata

Liliaceae
 Yucca harrimaniae

Linaceae
 Linum aristatum

Loasaceae
 Mentzelia albicaulis
 Mentzelia multiflora

Malvaceae
 Sphaeralcea coccinea
 Sphaeralcea parvifolia

Nyctaginaceae
 Abronia fragrans
 Abronia micranthus

Oleaceae
 Forestiera neomexicana
 Fraxinus anomala

Onagraceae
 Epilobium angustifolium
 Oenothera caespitosa
 Oenothera trichocalyx

Pinaceae
 *Pinus edulis

Plantaginaceae
 Plantago patagonica

Polemoniaceae
 Gilia aggregata
 Gilia gunnisoni
 Gilia leptomeria

Polygonaceae
 Eriogonum cernuum
 Eriogonum corymbosum
 Eriogonum leptocladon
 Eriogonum microthecum

Ranunculaceae
 Aquilega micrantha
 Clematis columbiana
 Delphinium scaposum

Rosaceae
 Amelanchier utahensis
 Cercocarpus intricatus
 Coleogyne ramosissima
 Cowania mexicana
 Rosa woodsii

Salicaceae
 Populus fremontii
 Salix amygdaloides

Santalaceae
 Comandra pallida

Scrophulariaceae
 Cordylanthus wrightii
 Penstemon comarrhenus
 Penstemon eatonii
 Penstemon utahensis

Umbelliferae
 Cymopterus fendleri
 Cymopterus purpureus

*Dominant

APPENDIX VIII
MODERN VEGETATION INVENTORY: UPLAND GRASS COMMUNITY — THE FLAT-LYING GRASS COMMUNITIES ABOVE THE CANYONS

Beverly J. Albee

Berberidaceae
 Berberis fremontii

Boraginaceae
 Cryptantha crassisepala
 Cryptantha flava
 Cryptantha jamesii
 Lappula redowskii
 Lithospermum incisum

Cactaceae
 Mammillaria tetrancistra
 Opuntia erinacea

Capparidaceae
 Cleome lutea

Compositae
 Artemisia carruthii
 Chaenactis stevioides
 Chrysothamnus nauseosus
 Gutierrezia sarothrae
 Haplopappus armerioides
 Haplopappus nuttallii
 Hymenopappus filifolius
 Hymenoxys acaulis
 Helianthus petiolaris
 Leucelene ericoides
 Malacothrix sonchoides
 Senecio longilobus
 Senecio multilobatus
 Senecio spartioides
 Townsendia incana
 Xylorhiza tortifolia

Cruciferae
 Descurainia richardsonii
 Erysimum asperum

Cruciferae (cont.)
 Lepidium montanum
 Stanleya pinnata
 Streptanthella longirostris

Cupressaceae
 **Juniperus osteosperma*

Elaeagnaceae
 Shepherdia rotundifolia

Ephedraceae
 **Ephedra nevadensis*
 Ephedra torreyana

Gentianaceae
 Frasera paniculata

Gramineae
 Aristida longiseta
 Festuca octoflora
 Hilaria jamesii
 **Muhlenbergia pungens*
 Oryzopsis hymenoides
 Stipa comata

Hydrophyllaceae
 Phacelia ivesiana

Leguminosae
 Astragalus ceramicus
 Astragalus mollissimus
 Astragalus musiensis
 Lupinus pusillus

Liliaceae
 Yucca harrimaniae

Linaceae
 Linum aristatum

Loasaceae
    Mentzelia albicaulis
    Mentzelia multiflora

Malvaceae
    Sphaeralcea coccinea
    Sphaeralcea parvifolia

Nyctaginaceae
    Abronia fragrans
    Abronia micranthus

Oleaceae
    Fraxinus anomala

Onagraceae
    Oenothera caespitosa
    Oenothera trichocalyx

Pinaceae
    **Pinus edulis*

Plantaginaceae
    Plantago patagonica

Polemoniaceae
    Gilia aggregata
    Gilia gunnisoni
    Gilia leptomeria
    Phlox austromontana

Polygonaceae
    Eriogonum cernuum
    Eriogonum corymbosum
    Eriogonum leptocladon
    Eriogonum microthecum

Ranunculaceae
    Delphinium scaposum

Rosaceae
    Cercocarpus intricatus
    Coleogyne ramosissima
    Cowania mexicana

Santalaceae
    Comandra pallida

Scrophulariaceae
    Cordylanthus wrightii
    Penstemon utahensis

Umbelliferae
    Cymopterus fendleri
    Cymopterus purpureus

*Dominant

APPENDIX IX
THE ANALYSIS OF HUMAN COPROLITES FROM
COWBOY CAVE

Patrick F. Hogan

INTRODUCTION

Beginning with Young's identification of
seeds in dried human feces found at
Salts Cave, Kentucky (Young 1910), coprolite studies have gradually become
accepted by American archeologists as
reliable indicators of prehistoric diet.
Since coprolites are the remains of
foods actually ingested, we can avoid
the uncertainty inherent in reconstruc-
tions of diet from floral and faunal
remains found at the site. However,
with the advantage of having a direct
index of food intake, the analyst must
accept the disadvantages imposed by the
nature of the data. The principal
problems inherent in such studies are:
1) A single coprolite sample represents
the remains of the food ingested by a
single individual over a 24 to 36 hour
period. Its contents are not only subject to the vagaries of individual preference and appetite, but are dependent
on the relative success or failure of
the food-gathering forays. 2) Material
within the fecal mass is a residue of
the digestive process. Relative frequencies of the components are, therefore, inversely proportional to the
nutrients assimilated by the body. Some
highly digestible materials may leave
little or no trace of their ingestion.
Despite the limitations, the study
of coprolites can make a significant
contribution to the reconstruction of

past subsistence patterns. Because of
these limitations, however, a procedure
of analysis that yields a maximum of information with a minimum input of man-
hours is indicated. In order to control
for the wide variation of components in
the individual specimens and illuminate
the overall similarities in diet, the
number of coprolites processed needs to
be sufficiently large to provide a reasonable estimate of the theoretical population. Since there are always limitations on the time and manpower available
for any given study, it seemed advanta-
geous to minimize the time devoted to
analyzing each sample, thereby permitting
more samples to be processed. Since the
complexities of the biological processes
that produced the coprolite allow us to
make only a rough estimate of the relative importance of the foods ingested,
involved and highly exact procedures of
quantification are not warranted. The
ideal procedure appeared to be one in
which identification and reasonably accurate estimates of relative frequencies
of components could be rapidly made on a
suite of samples selected to provide a
valid subsample of the coprolite population from the site.

Sample Size

Over 80 coprolites were recovered
during the excavation of Cowboy Cave;
due to the time required for experimen-
tation with analytical procedures,

however, it was decided that only about half this number could be processed [Hogan 1977]. Preliminary study of the stratigraphy and cultural materials suggested that the site could be divided into four units of occupation. These were taken as convenient groupings for stratigraphic sampling, since any changes in subsistence patterns through time would likely be reflected in other cultural changes. Since stratigraphic control during excavation was generally tight, and since each occupation level was sealed by essentially sterile sands, a stratum to be sampled was selected from each unit on the basis of having a maximum number of coprolites deposited over a relatively short period of time, the rate of deposition being indicated by the thickness of the deposit and the ladder of radiocarbon dates. All samples from Strata IIb and IVc were processed; the other two strata, IIIi and Vb, each contained so many coprolites that a 50 percent random sample was used in the analysis.

The use of stratigraphic sampling as described is seen as a significant improvement over simple random selection from the entire span of occupation. The sampling procedure provides four discrete samples, each of a temporal period estimated to be not longer than 200 to 300 years. Admittedly, the small sample size raises some question of the adequacy of the samples. However, since the site seems to have been a seasonally occupied, special-use area, and since the occupants were relating to a relatively harsh environment with a simple technology, it seems reasonable to expect that the diet was quite restricted and stable. Any change in resource utilization would have been triggered by environmental change or gradual technological innovation, and in such a case, the stratification should aid significantly in making the shift apparent.

Preliminary Processing

The initial problem in analyzing coprolites from an archeological site is the separation of human from non-human fecal material. During initial sorting,

the Cowboy Cave material was separated into three groups: probable human, herbivore, and carnivore. Although size and form provided some clues, the range of variation in human feces is so broad that these criteria were given little weight; the major basis for segregation was composition. After sorting, each coprolite was washed with a high-pressure stream of tap water to remove any extraneous material adhering to the outside. A second wash with distilled water was made to remove any pollen contaminates, and the coprolite was then split longitudinally. Half the sample was stored against the necessity of resampling or restudy; the second half was placed in a labeled jar that had been rinsed in distilled water. The jar was then filled nearly full with the reconstituting solution, 0.05 percent trisodium phosphate (Callen 1965:335), and capped.

Subsamples from each coprolite were then sent to the Composition Analysis Laboratory in Fort Collins, Colorado, for quantification of the tissue and identification of bone and hair. Results of their studies indicate that with the exception of some *Opuntia* spp. and a few traces of other epidermal tissues, all the tissue from the Cowboy Cave samples was endodermis, seed tissues, or other cellular plant material lacking the distinctive structure necessary for identification. The frequency given for the few identifiable tissues was expressed as a visual estimate of volume since the time required for normal quantification procedures was not warranted (Theresa Foppe, personal communication). No identifiable bone fragments were noted, and hair was identified only to the family. The frequencies of occurrence of seeds, along with a presence-absence notation of these identifiable components, then, were used to evaluate the diet of Cowboy Cave's aboriginal inhabitants. The relative frequencies of the seeds were expressed as a proportion of each taxa's frequency to the total seed count. Identification of taxa for computation purposes was made at the generic level except for the seeds of some unidentifiable grasses and the seeds of the families

Chenopodiaceae and Amaranthaceae. All grasses were classified as Gramineae, except *Sporobolus cryptandrus* which was counted separately. Chenopod and amaranth seeds are difficult to distinguish even when whole; therefore the fragments of these were classified as "Cheno-Ams" following a similar convention in palynology. With the exception of *Helianthus* spp., the seeds were nearly all the same size; consequently no volume corrections were made. Since the seed coats of the *Helianthus* fragmented into eight to ten pieces, the resulting overestimation of frequency was used as an informal compensation for the greater volume represented by the sunflower seeds.

The major technical problem encountered in obtaining the counts was that of assuring that the fragments counted were randomly selected. Each sample was prepared for counting by pouring the vial onto a glass sheet. After a thorough mixing, the sample was quartered-down following the method outlined by Krumbein and Pettijohn for subsampling geological sediments.

> . . . place four rectangular sheets of smooth paper, 2 x 4 in., together in such a way that each overlaps one half of one other and so that altogether they form a square. The [sample] should be carefully poured into the center of this square and flattened out into a circle. The pieces of paper may then be pulled apart, alternate quarters rejected, and alternate quarters combined and the process repeated as many times as necessary to give the desired size of sample. (Krumbein and Pettijohn 1938:357)

This procedure was modified to the extent that a spatula rather than the papers was used for quartering. This process, although seemingly elaborate, could be done rapidly and assured that a representative sample was placed under the microscope.

The next major technical problem involved insuring that the technician could not arbitrarily choose which elements to count and which to ignore. This was accomplished by placing a grid of 0.5 cm. squares under the viewing stage of the dissecting microscope. The subsample was spread as evenly as possible over the entire grid, and squares were randomly selected for counting. All of the fragments touched by a square were then counted before a second square was designated. The technician consequently had no control over which fragments were included in the count.

EVALUATION

As has been previously stated, the primary goal of analyzing the coprolites from Cowboy Cave was the illumination of prehistoric dietary patterns for the area. The immediate aims of the study were the recognition of the major food resources utilized, detection of any scheduling conflicts (that is, the harvesting of one resource at the expense of another), and the recognition of any changes in food resource usage through time. Two types of data available from the analysis could be applied to these questions: the proportions of coprolites containing a given component and the relative frequencies of each seed taxa within a single coprolite.

Given the form of the data, the three questions posed by the research goals may be rephrased as problems of pattern recognition, problems which can be attacked with statistical tools. The recognition of the major resources becomes a matter of noting which food items occur most frequently in the coprolite sample and which items occur with the greatest relative frequencies. For the Cowboy Cave material, proportions of coprolites containing a given component were calculated using all identified materials; tabulation of the relative frequencies of components, however, were limited to seeds. Because only a few samples were processed for each stratum, and because there was a wide variation in frequencies of components, data summaries using means and standard deviations are misleading. The most meaningful summary of component frequencies is the median, which accurately reflects the relative importance of any given component in a single stratum.

The proportions of coprolites containing each taxa (Table 1) suggest that the seeds of *Helianthus* and *Sporobolus* were the diet staples, with Cheno-Ams becoming important in IVc and Vb. Other seeds--*Dicoria*, *Corispermum*, and *Carex*-- and cactus pads appear to have been secondarily important in IIb and IIIi. The median values and frequency distributions, plotted by strata (Table 2) indicate that the staples identified are correct, but that the importance of the other taxa in IIb and IIIi were overestimated.

The second question posed by our research strategy, that of scheduling complement or conflict, is in fact, given the form of the data, a matter of covariation. Pearson product moment coefficients were therefore calculated for each pair of components, first utilizing the entire sample population, then by stratum. As indicated by the r^2 values in Table 3, there are no demonstrable significant covariations in seed components.

The final evaluative procedure addresses the problem of change in resource use through time. Such changes may be reflected in altered frequencies of resources, or the disappearance of formerly utilized food items. The first step in statistically examining the data for such patterns is to determine whether the food resources used in one stratum differ significantly from those used in other strata. The second step is to identify where the source of that variation lies. Discriminant analysis is a convenient method of doing both tasks simultaneously.

Briefly, discriminant analysis plots a point representing each coprolite sample in a multidimensional space, using the values of each component as vectors. The clusters formed by those points are compared with predesignated "groups" as defined by the investigator. In this

TABLE 1

PROPORTION OF SAMPLES CONTAINING

EACH ELEMENT

Element	Stratum			
	IIb	IIIi	IVc	Vb
Helianthus	.916	.5	.857	.454
Sporobolus	1.0	.736	1.0	.636
Cheno-Ams	.083	.333	.714	1.0
Corispermum	.333	.555	.142	.363
Dicoria	.666	.333	.142	.272
Carex	.416	.055	0.0	.090
Other Gramineae	.25	.277	.142	0.0
Opuntia	.416	.611	.285	.090

TABLE 2

MEDIAN VALUES AND FREQUENCY DISTRIBUTION

Stratum	*Helianthus*	*Sporobolus*	*Dicoria*	*Corispermum*	*Carex*	Cheno-Ams	Other Gramineae
IIb	.992	.935	.486	.047	.229	.001	.166
	.937	.848	.147	.020	.145	.000	.012
	.727	.830	.082	.004	.139	.000	.000
	.573	.670	.046	.003	.005	.000	.000
	.531	.393	.011	.000	.001	.000	.000
	→ .481	→ .392	→ .011	→ .000	→ .000	→ .000	→ .000
	.395	.362	.009	.000	.000	.000	.000
	.243	.354	.002	.000	.000	.000	.000
	.235	.190	.000	.000	.000	.000	.000
	.032	.069	.000	.000	.000	.000	.000
	.012	.022	.000	.000	.000	.000	.000
	.000	.005	.000	.000	.000	.000	.000
IIIi	.944	.971	.924	.966	.020	.928	.076
	.944	.954	.329	.227	.000	.650	.035
	.887	.945	.228	.028	.000	.453	.003
	.814	.590	.063	.023	.000	.045	.001
	.800	.538	.007	.007	.000	.045	.000
	.771	.404	.002	.004	.000	.013	.000
	.663	.200	.000	.002	.000	.003	.000
	.524	.123	.000	.002	.000	.000	.000
	→ .041	→ .093	→ .000	→ .000	→ .000	→ .000	→ .000
	.000	.076	.000	.000	.000	.000	.000
	.000	.066	.000	.000	.000	.000	.000
	.000	.041	.000	.000	.000	.000	.000

TABLE 2 (continued)

MEDIAN VALUES AND FREQUENCY DISTRIBUTION

Stratum	*Helianthus*	*Sporobolus*	*Dicoria*	*Corispermum*	*Carex*	Cheno-Ams	Other Gramineae
	.000	.033	.000	.000	.000	.000	.000
	.000	.014	.000	.000	.000	.000	.000
	.000	.007	.000	.000	.000	.000	.000
	.000	.000	.000	.000	.000	.000	.000
	.000	.000	.000	.000	.000	.000	.000
	.000	.000	.000	.000	.000	.000	.000
IVc	.639	.982	.014	.542	.000	.916	.001
	.363	.605	.000	.000	.000	.888	.000
	.362	.591	.000	.000	.000	.452	.000
	→ .062	→ .337	→ .000	→ .000	→ .000	→ .045	→ .000
	.011	.110	.000	.000	.000	.023	.000
	.001	.021	.000	.000	.000	.000	.000
	.000	.001	.000	.000	.000	.000	.000
Vb	.423	.176	.102	.682	.252	.998	.000
	.160	.161	.003	.260	.000	.911	.000
	.125	.118	.001	.109	.000	.888	.000
	.074	.110	.000	.001	.000	.833	.000
	.003	.092	.000	.000	.000	.788	.000
	→ .000	→ .087	→ .000	→ .000	→ .000	→ .757	→ .000
	.000	.001	.000	.000	.000	.739	.000
	.000	.000	.000	.000	.000	.678	.000
	.000	.000	.000	.000	.000	.524	.000
	.000	.000	.000	.000	.000	.318	.000
	.000	.000	.000	.000	.000	.297	.000

TABLE 3

r^2 OF SEED CATEGORIES
COMBINED SAMPLES OF ALL STRATA

	Helianthus	Sporobolus	Dicoria	Corispermum	Carex	Cheno-Ams	Other Gramineae
Helianthus	1.0	.063	.000	.063	.011	.260	.010
Sporobolus		1.0	.040	.043	.000	.183	.028
Dicoria			1.0	.009	.008	.048	.006
Corispermum				1.0	.000	.000	.006
Carex					1.0	.005	.004
Cheno-Ams						1.0	.027
Other Gramineae							1.0

case, the groups were defined on the basis of the stratum in which the sample occurred. The analysis procedure involves the selection of those variables (in this case dietary items) which most closely maintain the integrity of the groups as defined. Each sample is weighed according to these "discriminating functions," and the sample is then reassigned to the group with which it shares the greatest affinities. If the groups are quite distinct, all cases will be reassigned to their original grouping; if, however, there is a great deal of overlap in the points as plotted, a number of samples may be reassigned to another group. Since the goal of the program is to make the groups as distinct as possible, the percentage of cases reclassified is a good indicator of the integrity of the initially defined groups. In the same manner, the variables chosen as "discriminating functions" are those that show the greatest variation from one group to the next, and therefore can be interpreted as the dietary items varying most through time. Conversely, those important food items of little use in discriminating the groups can be viewed as stable, fairly constant resources.

Discriminant analysis on the Cowboy Cave material was done using the subprogram DISCRIMINANT of the SPSS package (Nie et al. 1975). The program was run on all components keyed to a presence-absence dichotomy, and on the seed frequencies alone. Selection of discriminating variables was done in a stepwise method, entering the best discriminating variable first, then the second best, and so on, until all variables had been entered or until the level of discrimination was not enhanced by the addition of more variables. The criteria for discriminating power was MINRESID, a selection of the variable that minimized the residual variation, thus tending to separate groups that were close together. Analysis indicated that separation of the groups was poor with a significant overlapping of values. The best separation of groups achieved was only an overall success in reclassification of 66.67 percent, with Stratum IIb, the most distinct group, having 83.3 percent of the cases correctly reclassified. Other strata had reclassification frequencies of 50 percent or less. These results indicate that rather than each stratum exhibiting a distinctive pattern of resource procurement, they collectively document a single procurement system gradually shifting to a new primary resource.

As indicated by the discriminating functions, there are two **major** sources of variation through time. The first is the presence and frequency of Cheno-Ams, the first function loaded in both runs. The second is the presence or absence of *Carex*, loaded as the second function in the seed run and as the third function in the analysis of all components. Other variables received insignificant discrimination scores or were not introduced into the analysis.

To summarize the results of the evaluation of the Cowboy Cave material, then, *Helianthus* and *Sporobolus* appear to have been the primary focus of the food-gathering activities in all levels except Vb, where Cheno-Ams were dominant. In Stratum IIb *Carex* was used moderately but it becomes insignificant thereafter. All other components were used to some extent throughout the time periods examined, but were always supplemental resources.

INTERPRETATION

The evidence supplied by the analysis of coprolite material from Cowboy Cave supports the interpretation of the site as a seasonal camp functioning primarily as a seed-processing station. With the exception of epidermis from cactus pads (*Opuntia* spp.), the bulk of the identifiable plant materials in the coprolites was seed fragments. Though meat consumption cannot be judged directly from the coprolites, the near lack of odor during the reconstitution suggests that only small quantities of

skatole and **endole** were present. These compounds are products from the putrification process in the intestine that results largely from the breakdown of proteins; consequently, their absence suggests a largely vegetable diet. Some bone and hair indicates that rabbits (Lagomorpha) and rodents (Rodentia) were occasionally procured to supplement the diet. Bone scrap and miscellaneous pieces of fur found within the deposits in the later occupation periods (see Lucius and Hull, this volume) suggest the supplemental hunting of big game as well.

The site appears to have been periodically occupied during the late summer and early fall (mid-August through September), judging from the time that the seeds important in the diet at this site ripen (see Table 4).

The staple food resources in the early occupations (8300 B.P. to 6700 B.P.) were sunflower (*Helianthus annomolus*) and sand dropseed (*Sporobolus cryptandrus*). Lesser use was made of dicoria (*Dicoria brandegei*), bugseed (*Corispermum hyssipfolium*) and carex (*Carex* spp.). These latter seeds and the cactus pads were all found in a large proportion of the coprolites from these levels, but generally occurred only in small percentages. This suggests that they were gathered when chanced upon, but that the main business of the camp was the processing of sunflower and dropseed.

The *Helianthus* sp. seed identified in the coprolites was unroasted and frequently whole, with the hulls present but split away. It appears that the seeds were consumed hull and all.

Sporobolus spp. seeds appear to have been utilized both as meal and as whole seed. Bits of milling stone in a few of the coprolites appear to have been dislodged from the grinding surface and then been ground to a capsule shape during the back-and-forth milling process. Charcoal in the scats suggest that the seeds were sometimes roasted after winnowing. This parallels Steward's account of the Piutes processing *Oryzopsis* spp. in the Owens Valley (Steward 1933:244).

Sporobolus would probably have been

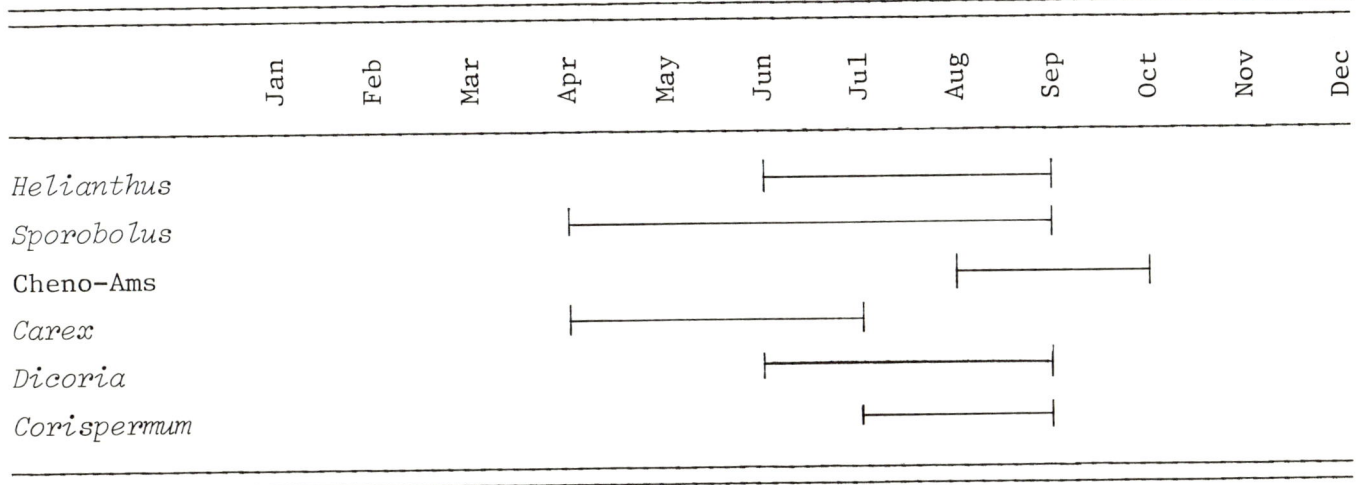

TABLE 4

FLOWERING TIMES FOR SEED TAXA RECOVERED
FROM COWBOY CAVE COPROLITES*

	Jan	Feb	Mar	Apr	May	Jun	Jul	Aug	Sep	Oct	Nov	Dec
Helianthus						├─────────────────────┤						
Sporobolus				├───────────────────────────┤								
Cheno-Ams								├────────────────┤				
Carex				├─────────────────────┤								
Dicoria						├─────────────────────┤						
Corispermum							├───────────────┤					

*Note: Seeds would ripen in the latter portion of the flowering period.

abundant on the rocky slopes and sandy, open uplands surrounding the cave. Sunflower, dicoria, and bugseed are weedy plants and would be expected to occur initially in such naturally disturbed areas as the bottoms of dry washes. The resulting situation of localized abundance would be well suited to the intensive harvesting techniques postulated for Cowboy Cave. Further, as man's activities disturbed other plant growth, new areas would become available for the establishment of these weedy communities.

The use of *Carex* suggested by the coprolites in Stratum IIb argues that a wetland environment was also close at hand during this early period. Most likely there was a perennial seep in the wash below the cave with enough flow to permit *Carex* to remain established. With the gradual deterioration of the environment the flow was diminished and that habitat destroyed.

From the quantity of ground stone found in the cultural deposits of Cowboy Cave and the lenses of chaff, it would appear that large quantities of seeds were gathered and then brought back to the cave, where they were winnowed,

sometimes roasted, and then ground into meal. It seems likely that such large quantities of seeds were processed for storage and later use. Since the cultural deposits are not extensive and the array of artifacts is not indicative of the range of activities to be expected at a base camp, it is surmised that the seeds were transported to some more permanent winter habitation.

Beginning about 6700 B.P. Cheno-Ams gradually emerge as an important dietary component. (While no distinction was made formally between *Chenopodium* and *Amaranthus* for reasons previously mentioned, my impression is that at Cowboy Cave *Chenopodium* was the more important resource since most of the seeds that could be identified were of that taxa). Initially, Cheno-Ams were but one more of the broad spectrum of weedy seeds utilized, but by 3600 B.P., they are codominant with sunflower and dropseed, and in Stratum Vb (1900 B.P.), they are the single most important seed resource.

The processing of Cheno-Ams at Cowboy Cave appears from the evidence to have been similar to that already described for dropseed, except that the Cheno-Am seeds seem to have always been ground prior to consumption. Again, based on the lack of extensive cultural deposits and the limited range of artifacts, it seems likely that the stay at the caves was just long enough to complete the seed gathering. The numerous cache pits in Units IV and V were most likely temporary storage facilities used as the seeds were being processed. After the work had been completed, the pits were cleaned out, and the seeds were transported to a winter camp. Why this gradual shift in the harvesting of wild seeds should occur is difficult to determine, especially in the narrow context presented at Cowboy Cave. One explanation is suggested by the fact that an emphasis on Cheno-Ams necessarily means a later scheduling of the wild seed harvest.

The seeds of both *Chenopodium* and *Amaranthus* tend to ripen later than dropseed or *Helianthus annomolus*. Further, in those coprolites where Cheno-Ams are dominant, the species of sunflower present is *Helianthus petiolaris* that blooms somewhat later than *H. annomolus*, and consequently, the seeds ripen later, i.e., in late September and October.

It is unlikely that this change to a period of utilization later in the growing season would occur for any trivial reason. As Flannery (1968:67) has indicated, hunter-gatherers are of necessity sophisticated lay botanists. In the unyielding environs of the canyon-lands regions, scheduling to stockpile winter food resources would not be a matter of whim. If, in the later occupations of the cave, there is a shift to a new resource that can be obtained later in the year but only at the expense of previously depended upon staples, there is bound to be a good reason for it. I suggest that the increasing emphasis on Cheno-Ams arose as a result of a scheduling conflict between the gathering of sunflower and dropseed, and the harvesting of maize. If crops were being grown elsewhere, time demands during the critical time of ripening and harvest would probably work against

scheduling any time for wild seed gathering during August or early September. After the harvest, however, some time would be available for such activities; and if the maize harvest were poor, such gathering expeditions would be a necessity. By shifting the emphasis of wild seed gathering to Cheno-Ams rather than *Sporobolus* and *Helianthus*, the scheduling conflict would have been resolved, and important additional stockpiles would have become available. While no evidence of corn was found within Unit IV at Cowboy Cave, it is generally acknowledged that maize and squash were grown on a limited scale in the Southwest by 2000 B.C. (Reed 1964; Whalen 1973), some 300 years earlier than the date for this unit.

In Unit V, the presence of corn pollen and macrofossils indicates such a scheduling conflict certainly existed. Since no corn was found in any of the coprolites it seems unlikely that crops were grown at Cowboy Cave; the corn that was present was probably brought in from elsewhere, though this is far from proven. Despite the unanswered questions, Cowboy Cave provides an intriguingly detailed picture of one procurement system and its change through time. That system's articulation with the overall economic system of the prehistoric peoples that camped there will remain unknown, however, until **broad-scale** investigations of the area can be conducted.

REFERENCES

CALLEN, ERIC O.

1965   Food Habits of Some Pre-Columbian Mexican Indians. *Economic Botany*, Vol. 19, No. 4, pp. 335-43. New York.

1967   Analysis of the Tehuacan Coprolites. In *The Prehistory of the Tehuacan Valley, Vol. 1: Environment and Subsistence*, Douglas S. Byers (Ed.). University of Texas Press, Austin.

CALLEN, ERIC O., AND T. W. M. CAMERON

  1960   A Prehistoric Diet Revealed in
          Coprolites. *The New Scientist*,
          Vol. 8, pp. 35-40. London.

FLANNERY, KENT V.

  1968   Archaeological Systems Theory
          and Early Mesoamerica. In *An-
          thropological Archaeology in
          the Americas*, Betty J. Meggers
          (Ed.). The Anthropological
          Society of Washington, Washing-
          ton, D. C.

FRY, GARY F.

  1967   Analysis of Prehistoric Copro-
          lites from Utah. *University of
          Utah Anthropological Papers*,
          No. 97. Salt Lake City.

HOGAN, PATRICK

  1977   The Analysis of Human Copro-
          lites From Cowboy Cave. Manu-
          script on file, University of
          Utah Department of Anthropology,
          Salt Lake City.

KRUMBEIN, WILLIAM C., AND F. J.
PETTIJOHN

  1938   *Manual of Sedimentary Petrology.*
          Appleton-Century-Crofts, Inc.,
          New York.

MARTIN, ALEXANDER C., AND LEROY J.
KORSCHGEN

  1963   Food Habit Procedures. In *Wild-
          life Investigational Procedures*,
          H. S. Mosby (Ed.). Edwards
          Brothers, Ann Arbor.

NIE, NORMAN H., ET AL.

  1975   *SPSS: Statistical Package for
          the Social Sciences.* McGraw-
          Hill Book Company, New York.

REED, ERIC K.

  1964   The Greater Southwest. In *Pre-
          historic Man in the New World*,
          Jesse D. Jennings and Edward
          Norbeck (Eds.), pp. 175-92.

The University of Chicago Press,
Chicago.

SPARKS, DONNIE R., AND JOHN C. MALECHEK

  1968   Estimating Percentage Dry Weights
          in Diets Using a Microscopic
          Technique. *Journal of Range Man-
          agement* Vol. 24, No. 4, pp. 261-
          65.

STEWARD, JULIAN H.

  1933   Ethnography of the Owens Valley
          Piute. *University of California
          Publications in American Archae-
          ology and Ethnology*, Vol. 33,
          No. 3. Berkeley.

STEWART, D. R. M.

  1967   Analysis of Plant Epidermis in
          Faeces: A Technique for Studying
          the Food Preferences of Grazing
          Herbivores. *Journal of Applied
          Ecology*, Vol. 4, No. 1, pp. 83-
          111. Oxford.

WHALEN, NORMAN M.

  1973   Agriculture and the Cochise.
          The Kiva, Vol. 39, No. 1, pp.
          89-96.

YARNELL, RICHARD A.

  1969   Contents of Human Paleofaces.
          In "The Prehistory of Salts
          Cave, Kentucky," P. J. Watson.
          *Illinois State Museum Reports
          of Investigations* No. 16, pp.
          41-5. Springfield.

YOUNG, BENNETT H.

  1910   The Prehistoric Men of Kentucky.
          Filson Club Publications No. 25.
          Louisville.

APPENDIX X
POLLEN ANALYSIS OF COWBOY CAVE CULTURAL DEPOSITS

La Mar W. Lindsay

INTRODUCTION

Pollen analysis of Cowboy Cave cultural deposits provides a spectrum of relative pollen frequencies that essentially spans the Altithermal (Antevs 1948, 1955) and encompasses subsequent Neoglacial events (Denton and Porter 1970). However, since the dating and exact nature of these events vary for specific locales (Bryan and Gruhn 1964) and because no natural pollen records exist for the central Colorado Plateau, inferences necessarily remain to be substantiated with a local natural pollen record. Still, the cave cultural record can be interpreted with the natural sequence from Snowbird Bog in the Wasatch Mountains (Madsen and Currey 1977), partial records from Sudden and Pint-Size shelters (Currey 1976a, 1976b; Lindsay 1976), the precultural record from Cowboy Cave (Spaulding and Petersen, *Appendix II*), and several natural sequences to the south on the Plateau and in the Great Basin (e.g., Petersen and Mehringer 1976; Mehringer and Warren 1976; Mehringer, Martin, and Haynes 1967). Together these records shed some light on the changing environment during man's occupancy of the cave.

Pollen identified from the modern surface sample generally reflects the descriptions of contemporary vegetation (Albee, *Appendices VII* and *VIII*). Cowboy Cave is in a mid-latitude desert/steppe, within the lower limits of the pygmy conifers. *Juniperus osteosperma* (Utah juniper) and *Pinus edulis* (double-leaf pinyon) are dominant both in the canyons and on the grassland flat above the cave. The latter consists of a variety of grasses, shrubs, and forbes of which *Muhlenbergia pungens* (scratchgrass), *Gutierrezia sarothrae* (snakeweed), and *Ephedra nevadensis* (Mormon tea) predominate. A variety of species are found in proximity to the cave including *Artemisia tridentata* (big sage), *Quercus gambelii* (Gambel oak), *Berberis fremontii* (barberry), *Amelanchier utahensis* (serviceberry), and *Cercocarpus intricatus* (mountain mahogany). Mesic species, including *Populus fremontii* (cottonwood) and *Salix* sp. (willow), are present in the vicinity. Although the identification of pollen at the species level is generally limited, macrofossil identifications (Barnett and Coulam, *Plant Macrofossil Analysis*) indicate that most extant species are represented in the cultural record. Also, the variety of unidentified grass species in most pollen samples, including the modern surface, is consistent with the large number in the canyon and on the flats above.

The principle value of pollen analysis of cultural deposits is to determine 1) the presence or absence of genera, 2) the significant changes in the relative proportions of various economic types such as grasses and the chenopods, and 3) the point at which *Zea mays* makes its appearance in the record. However, assuming that sampling error and distortion due to differential pollen preservation are minimal, a case may be made for at least a rough correspondence between changes in availability and use of various flora.

PROCEDURE

Twenty-eight soil samples, predominantly eolian, and representing precultural Stratum IIa (8690±75 B.P.) and cultural/depositional Strata IIb through Vb (8275±80-1890±65 B.P.), were selected for pollen analysis from a large number obtained during the 1975 field season. Two columns, where the strata were generally most abundant, were sampled in arbitrary increments from bottom to top. Approximately 300 g. of sediment per sample was collected. Seventeen soil samples (Strata IIa-IIIi), at a depth of 0.50 to 1.30 m. below the modern surface, were from a cross section profile about 6 m. north of a central point at the front of the cave. Samples IIIb, c, h, j, and k were not represented at this location. Nine samples (IVa-d), 0.05 to 0.45 m. below the surface, were from a nearby vertical cut about 7 m. north-northwest of the cave front. In addition, two samples (Va and b) that were unavailable during the original sampling were collected about 11.5 m. to the north-northwest. No samples were run from the uppermost strata (Vc and d). A modern surface sample was collected from the flat opposite and above the cave.

Nine metates representing various strata in the sequence, including one from Vd, were sampled for residual pollen. The grinding surfaces were dry brushed to remove matrix deposits, then washed and wire brushed using water obtained from Frenches Spring, about 13 km. distant from the cave. Eight coprolite samples were selected from Strata IVc (3635±55 B.P.) and Vb (1890±65 B.P.). The sampling was restricted to the two strata in order to concentrate on the variation between coprolites in each unit. This aspect of the pollen study, in conjunction with the macrofossil analysis (Hogan, *Appendix X*), was undertaken to provide definition of diet before and after the advent of plant domestication.

All samples were extracted following the procedures in general use for alluvial and eolian samples from the Southwest and the Great Basin (Mehringer 1967). Extracted material was stained with basic fuschin, mounted in glycerol and counted under 600X magnification. Identifications were made using modern reference material provided by Garrett Herbarium, University of Utah.

Pollen abundance and state of preservation varied between samples, with only a slight relationship to age. Most samples contained pollen that was heavily eroded and fragmentary. Poor preservation was particularly noticeable in the earlier metate samples and the Stratum Vb coprolites. Only 100 grain counts were obtained for many of these samples, while counts of 200 or more were obtained for the remainder, including those from the deposits. The coefficient of reliability for 100 grain counts is less than 0.90, which is ordinarily insufficient for inference. However, when several such samples are averaged their reliability is deemed acceptable (Barkley 1934). Five depositional samples (IIIe, and the sequence from IVc at 0.20 m. below the surface to IVd at 0.05 m.) contained little or no pollen.

THE POLLEN RECORD

The Snowbird Bog pollen record of the central Wasatch Mountains indicates that deglaciation following the Late Wisconsin had begun by 12,300 B.P. (Madsen and Currey 1977). Pollen analysis of basal precultural deposition (ca. 13,000 B.P.) at Cowboy Cave indicates Holocene warming was well underway, but that montane vegetation persisted in this locale to about 8700 B.P. (Spaulding and Petersen, *Appendix II*). The latter date marks the beginning of a rise in the importance of xeric vegetation at Cowboy Cave. Comparison of pollen samples from Unit II with the modern surface sample (Fig. 1) suggests that vegetation at this time was probably very similar to that of the present.

Initial cave occupancy occurred at 8275±80 B.P. Although some grass usage is indicated from the metate sample associated with Stratum IIb (Fig. 4), there is little other indication of man's introduction of economic flora into the cave. The similar pollen percentages for the precultural and cultural samples from Unit II and, subsequently, Stratum IIIa suggests that man had very little effect on the environment.

The nature of the Altithermal at Cowboy consisted of a gradual decrease in conifers with a corresponding increase in nonarboreal species (Fig. 2). The ratio of pinyon/juniper to *Artemisia* decreased to approximately half the 8500 B.P. proportions by ca. 6500 B.P. (Fig. 3). Grasses show a general increase throughout the entire period. This is probably the result of increased availability, but could also represent heavier human utilization with time. It may also reflect locally increasing populations occupying the cave.

This mid-postglacial dry period is essentially duplicated in various pollen records throughout the Great Basin and Colorado Plateau. Bright (1966) dates the interval from 8400 to 3100 B.P. in the extreme northeastern Great Basin, while Madsen and Currey (1977) date the event from 8000 to 5200 B.P. in the Wasatch Mountains of the eastern Great Basin. To the south, the Altithermal is dated from 8500 to about 4000 B.P. in the La Plata Mountains of southwestern Colorado (Petersen and Mehringer 1976) and although Martin (1963) has argued for a "wet interval" in the form of increased summer moisture derived from the Gulf, similar dating of climatic change is obtained from the various southern Arizona records.

The dating of the onset of the Altithermal seems fairly fixed at 8500–8000 B.P. throughout the region and, with the exception of the data from the La Plata Mountains and southern Arizona, a ca.

5000 B.P. climatic reversal—the Neoglacial—is evident in most records. This event is dated between 5500 and 5000 B.P. from the White Mountain, California tree-ring record in the western Great Basin (LaMarche 1974), 5000 B.P. in the San Juan Mountains, New Mexico (Andrews et al. 1975), and although not specifically identified as such, the event may also be seen in a pollen record from the Amargosa Desert (Mehringer and Warren 1976). More mesic conditions (and cooler?) occur ca. 5000 B.P. at Murray Springs in southern Arizona (Mehringer, Martin, and Haynes 1967), but it is difficult to equate this with Neoglacial cooling because of the diametric weather systems that contribute to cooler and to wetter conditions (Aschmann 1958). However, Petersen and Mehringer (1976) suggest cooler temperatures prevailed by ca. 4000 B.P. in the La Plata Mountains of the southeastern Plateau.

The onset of the Neoglacial in the northeastern Great Basin is dated at 5200 B.P. in the northern Wasatch Mountains (Madsen and Currey 1977). At Sudden Shelter, 130 miles to the south on the Wasatch Plateau and 70 miles northwest of Cowboy Cave, the abrupt increase in *Pinus* and the slight to moderate increases in fir and spruce antedating 4900 B.P. (Lindsay 1976) may well be associated with Neoglacial cooling rather than the tentative suggestion that the increases in arboreal pollen and concomitant rates of colluviation are the product of Mexican monsoons (Currey 1976b; Lindsay 1976).

A Neoglacial event is postulated at Cowboy Cave, occurring about mid-point between Strata IIIi (6390±70 B.P.) and IVc (3635±55 B.P.). The doubling of *Pinus* pollen percentages associated with Stratum IVa and the lower portions of IVb, and the simultaneous occurrence at Sudden Shelter suggests a ca. 5000 B.P. dating of the event. With the deposition of Strata IVb the ratio of conifers to *Artemisia* increases seven times over Altithermal levels. This suggests that forest conditions in the Cowboy locale were considerably more extensive than at present. Generally, a greater variety

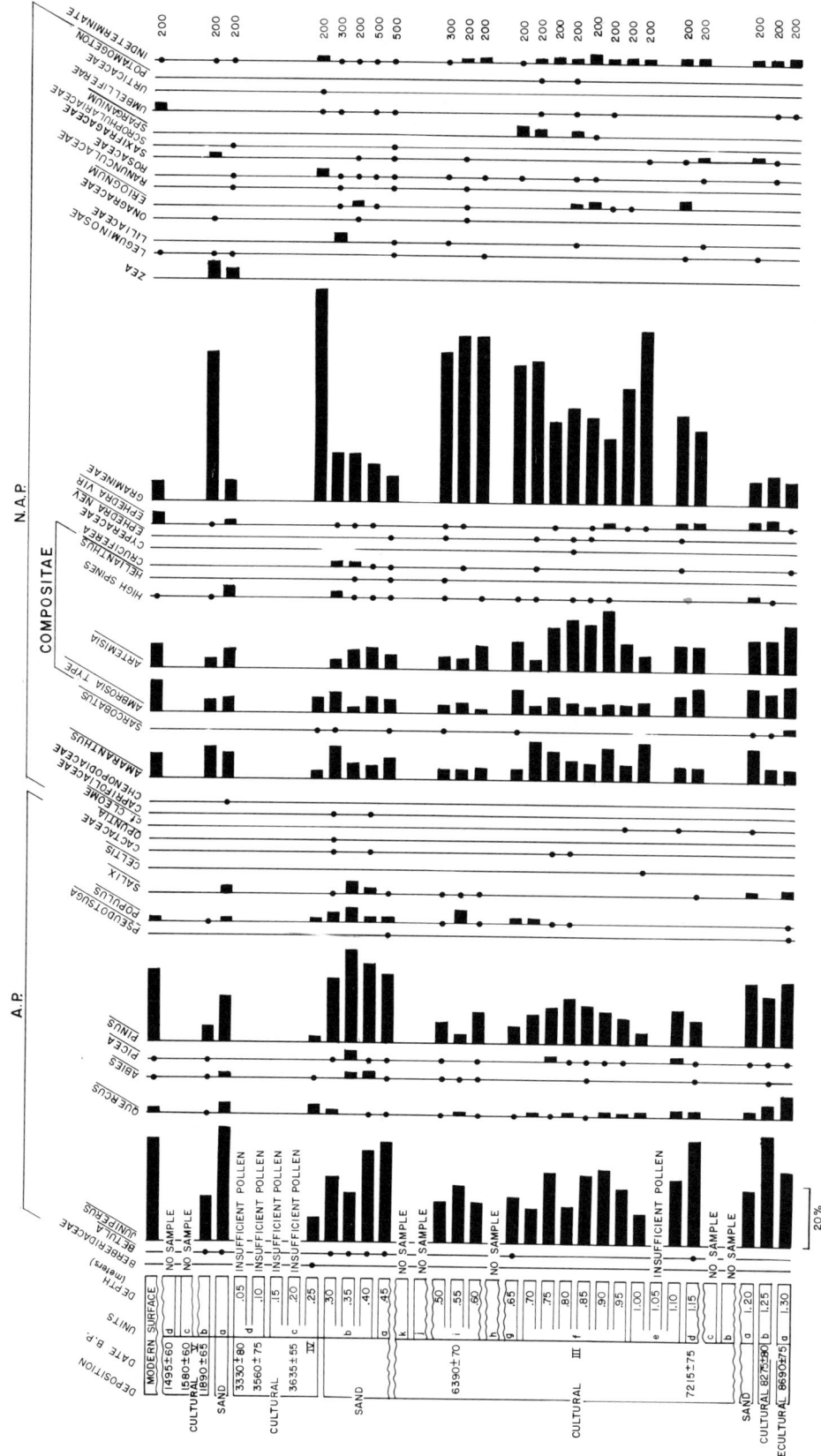

FIG. 1. Pollen diagram--relative percentages from Cowboy Cave cultural deposits

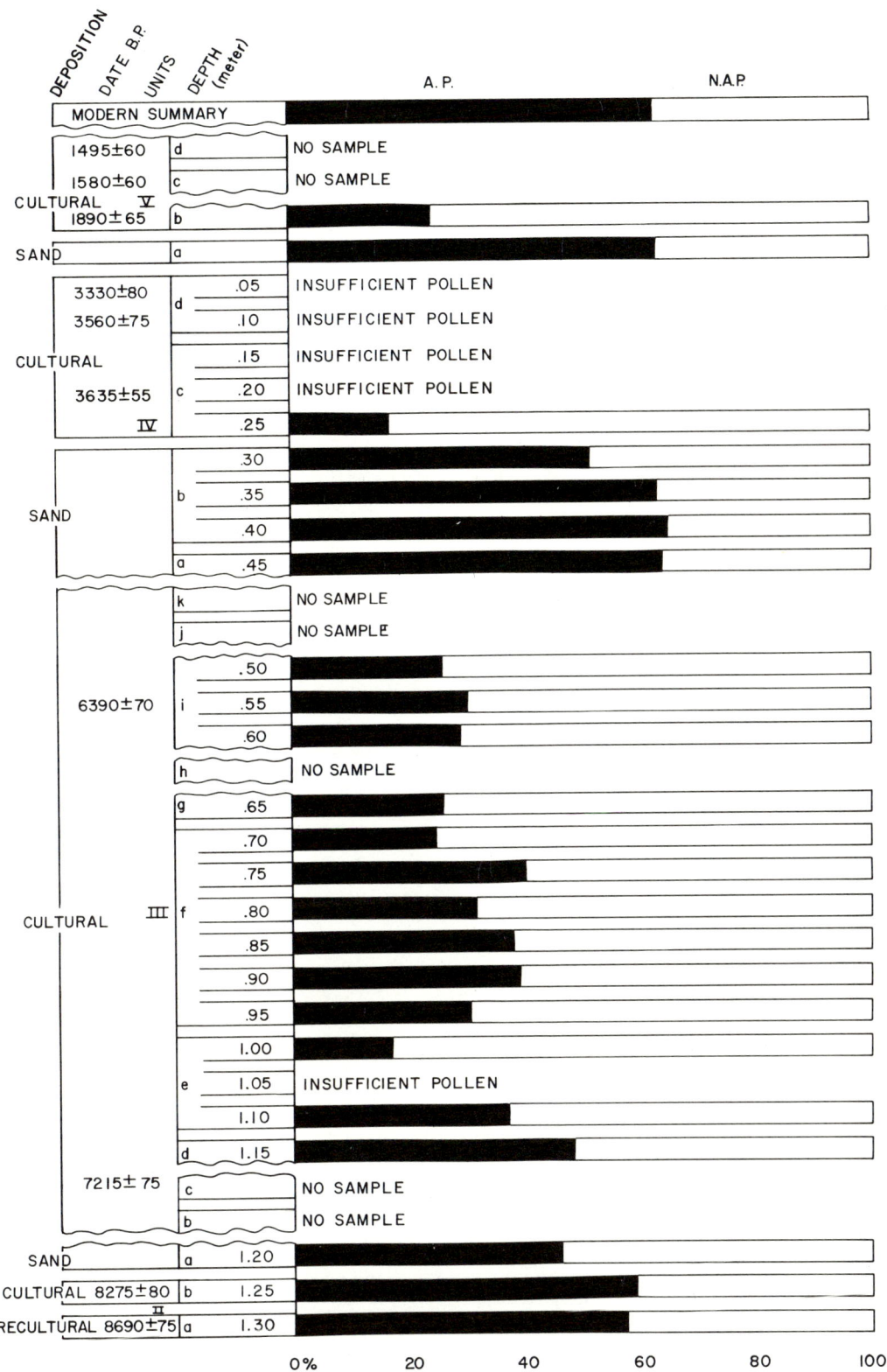

FIG. 2. Arboreal/nonarboreal pollen summary

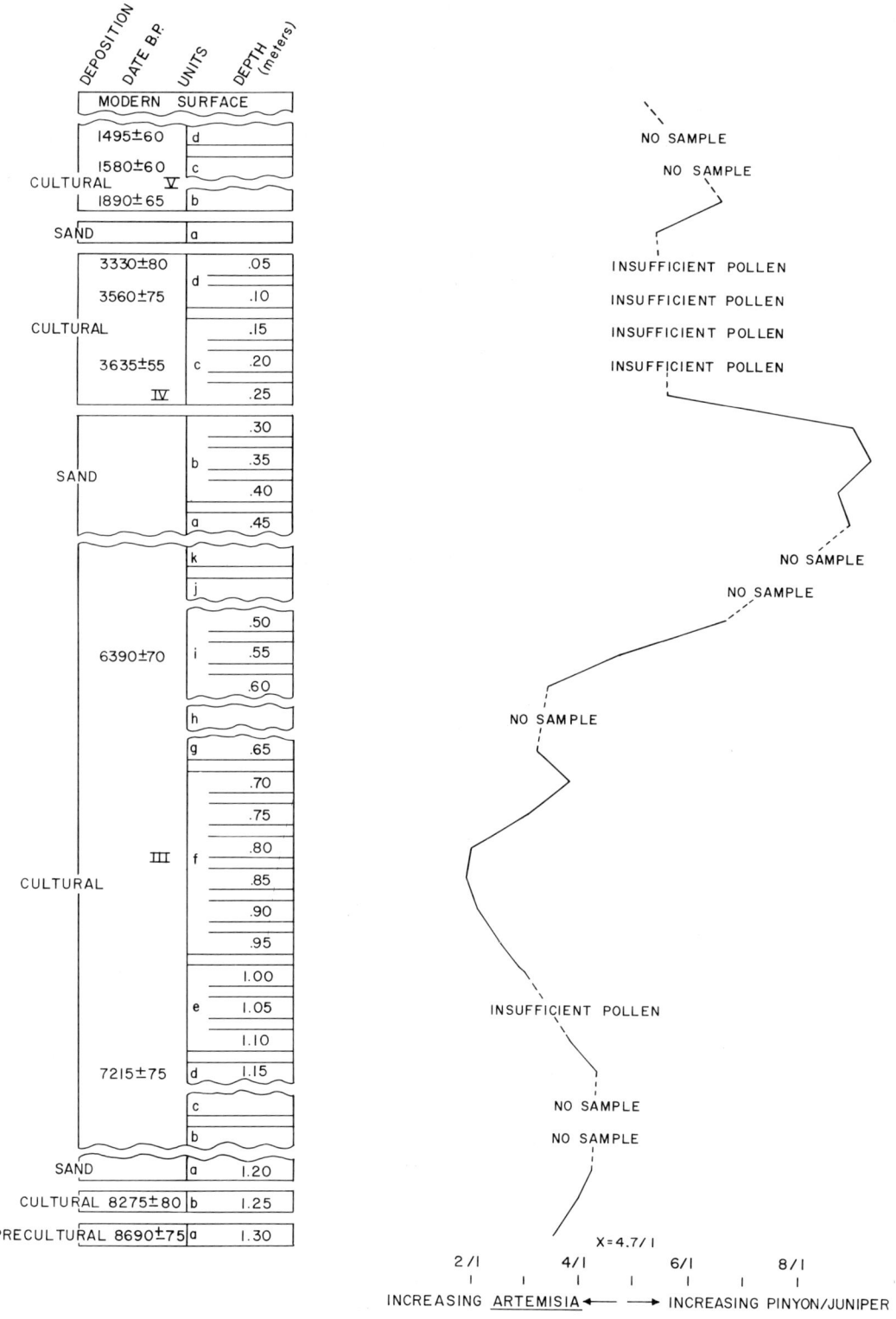

FIG. 3. Pinyon/juniper divided by *Artemisia* ratio (The curve has been smoothed by a weighted, three level moving average $\frac{a + Bb = c}{4} = X$)

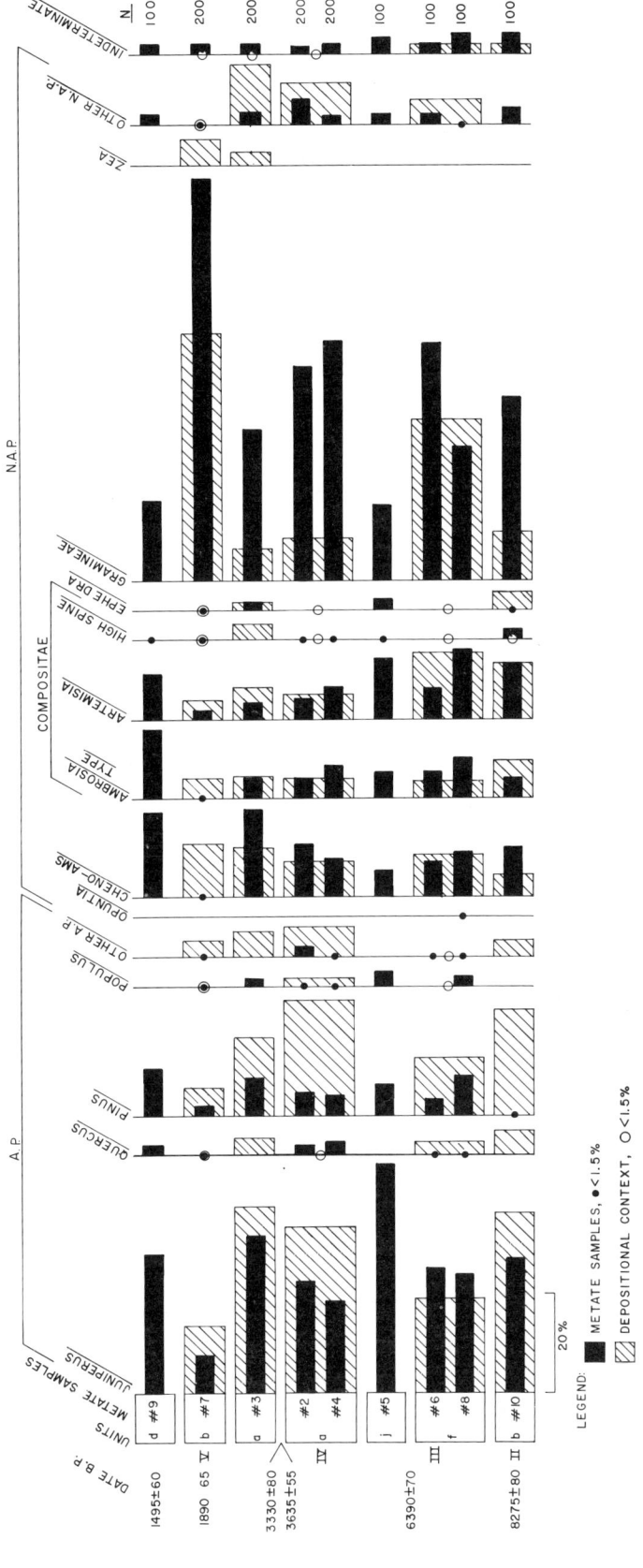

FIG. 4. Pollen diagram--relative percentages of selected metate samples and their respective depositional contexts

of pollen types occur with the increased forest, but not in appreciable amounts. The latter is probably due to the constraints imposed on the representation of nonarboreal species by the higher conifer percentages. The cave was apparently unoccupied during much of the time, although two metate samples (Nos. 2 and 4), with proveniences of Stratum IVa, were probably derived from the basal portion of Stratum IVb. The metate samples show extremely high grass usage.

This apparently short-lived early Neoglacial stadial ended prior to the deposition of the basal sample of Stratum IVc and prior to reoccupation of the cave. Currey (1976a) places the end of the early stadial about 4500 B.P., and this agrees closely with the 4600 B.P. date of an abrupt decrease in arboreal pollen (except *Juniperus*) at Sudden Shelter (Lindsay 1976).

With the onset of cave occupancy during the early/mid-interstadial, high grass percentages were obtained for both the depositional sample IVc (0.25 m.) and the 3635±55 B.P. coprolite samples (Fig. 5). However, the latter are not significantly greater than the context from which they were derived. It is unclear why remaining samples from Stratum IVc and those from IVd (3635±55-3330±80 B.P.) are, with the exception of a few highly eroded and fragmented grains, entirely devoid of pollen. The results of sediment analysis are almost entirely complacent throughout all deposits (Currey 1978, personal communication), however Strata IVc and d were not included in the investigation. Also, a rough calculation of deposition rates suggests little change over time. The absence of pollen in samples spanning a 300 year interval almost certainly reflects poor preservation.

A second or mid-Neoglacial stadial is postulated (ca. 3300 B.P.) for the region (Currey 1976a). This is likely represented at Cowboy (possibly slightly later) in the moderate *Pinus* and high juniper percentages associated with sample Va. Once again, the cave was apparently abandoned during **a cooling**

interval. *Zea mays* first appears in the pollen record in the Va sample, clearly derived from the interface with Stratum Vb, when the cave was again reoccupied. *Zea* constitutes about five percent of the latter sample. This marks the advent of domesticated plants at Cowboy Cave at or just prior to 1890±65 B.P. The Unit V metate samples were entirely devoid of corn pollen, and only token amounts (> 1.5 percent) were identified in the Vb coprolites. Other pollen identified in Unit V deposits is not significant, and no samples were available for the uppermost Strata, Vc (1580±60) and d (1495±60).

In sum, Cowboy Cave was occupied at the onset of the Altithermal (ca. 8500 B.P.) and either abandoned or used rarely during the subsequent early stadial of the Neoglacial (ca. 5000-4500 B.P.). The development of the pinyon/juniper forest, which was reduced to well below that of the present day during the Altithermal, apparently increased significantly during the subsequent Neoglacial. Forest fluctuations probably continued with subsequent Neoglacial events; however, insufficient sampling and limited pollen preservation preclude exact estimates. The cave was subsequently reoccupied before 3600 B.P., during at least the latter portion of the early/mid-interstadial (ca. 4500-3300 B.P.), and again abandoned with the onset of the mid-Neoglacial. The final occupation occurs prior to ca. 1900 B.P. during drier conditions and coincides with the appearance of *Zea mays* in the record. Grasses show a general and gradual increase over the full duration of intermittent cave use while *Artemisia* and *Quercus* correspondingly decrease. Other pollen are unrevealing except to note that Cheno-Ams, the Compositae and other potential economic plants were available in varying proportions throughout the record. The presence of *Populus* during the mid-Altithermal suggests that water was available during even the driest periods.

THE ECONOMIC POLLEN RECORD

The pollen samples obtained from the

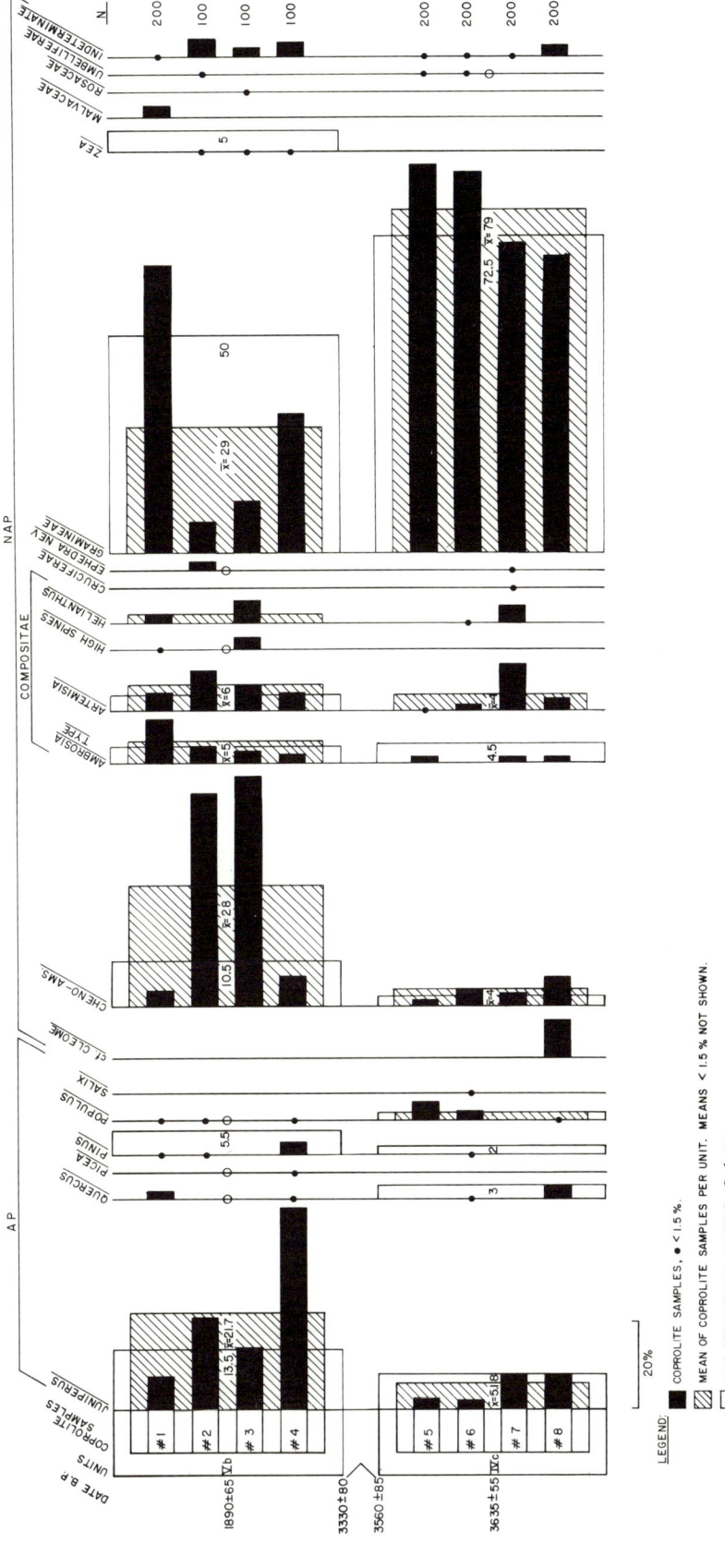

FIG. 5. Pollen diagram—relative percentages of coprolite samples from units IVc and Vb

metates (Fig. 4) and coprolites (Fig. 5), taken with the coprolite macrofossil analysis (Hogan, *Appendix X*), indicate a subsistence shift with the final period of cave occupancy. Since identification of most pollen types is limited to genera, and the various grasses are indistinguishable, reference to flora at the species level is by necessity deferred to the macrofossil analyses (Barnett and Coulam, *Plant Macrofossil Analysis*; Hogan, *Appendix X*).

All metate samples but No. 8 (Stratum IIIf) show high grass usage when compared to the depositional samples from which they were derived. The grass counts for the samples from IIIj (No. 5) and Vd (No. 9) are somewhat low and, unfortunately, the depositional samples are not available for comparison. The high grass counts, based on the metate samples alone, indicate an extremely heavy reliance on grass at all times of cave occupancy. This is well supported in both the pollen and the macrofossil analysis of the coprolites. Grit from the metates was noted in the coprolite samples from Strata IIb and IIIi (Hogan, *Appendix X*).

Other pollen from the samples is less revealing. Juniper counts are relatively high, but not significantly greater, and in several cases, much less than those obtained from the deposits. Cheno-Ams exceed deposit counts in IVa and Va, and only the latter sample suggests possible utilization. The remaining pollen, such as *Quercus*, *Populus*, and the Compositae, while possibly used, are not particularly informative. Also, the Unit V metate samples are entirely devoid of corn pollen. The limited counts obtained from the earlier samples reflect poor pollen preservation. Why this trend is not seen in the depositional samples is unknown.

Consistent with the pollen counts obtained from the metates, high grass usage is also evident in many of the coprolite samples. The extremely high percentages of grass pollen obtained from the Stratum IVc samples, despite the little difference from that obtained from the deposit, leaves little

doubt regarding the heavy reliance on grass. The tentative identification of *Cleome* (No. 8), and the *Artemisia* and *Helianthus* counts (No. 7) show these plants were also utilized. *Cleome* (beeweed) was identified as a major dietary component of the Anasazi in the Glen Canyon (Martin and Sharrock 1964). The remaining types and *Populus* are variously recorded in the ethnobotanical record (e.g., Chamberlain 1911; Steward 1938; Kelly 1964; Whiting 1939).

Some economic diversity, which coincides with the appearance of *Zea*, is suggested in the Stratum Vb coprolites. Grass is heavily represented in two of the samples. Cheno-Ams are almost three times the Vb deposit count. *Juniperus* is extremely high in one sample (No. 4) and moderately represented in another (No. 2). *Helianthus* and other Compositae appear to have served at least as a portion of the diet. *Zea* is represented in only token amounts (> 1.5 percent). Why pollen preservation is poor in the Vb samples is unknown.

Most revealing is the comparison between means of the Strata IVc (ca. 3600 B.P.) and Vb (ca. 1900 B.P.) samples. Grass usage appears to have shifted from about 80 to 30 percent while availability decreased less than 25 percent. Cheno-Ams shifted from about 5 to more than 25 percent, while availability increased less than 10 percent. Juniper shifted from 5 to about 20 percent while availability remained essentially unchanged. These differences point to a major subsistence shift by 1890±65 B.P., which coincides with the appearance of corn at the cave.

Hogan (*Appendix X*) has postulated that the introduction of horticulture may have dictated a change in scheduling, because Cheno-Ams, available later in the season, appear as a dominant food source in the Cowboy Cave record. The subsistence shift may also be due to the newly introduced, somewhat fortuitous, and unpredictable corn yield that dictated a heavier reliance on a greater variety of species.

SUMMARY AND CONCLUSIONS

Grasses were most abundant at Cowboy Cave during the Altithermal (ca. 8500-5000 B.P.) and the subsequent Neoglacial interstadial (ca. 4500-3300 B.P.). During the intervening periods of cooling, which promoted forest development, the cave was apparently abandoned.

Corn appears in the record at Cowboy with the suggested onset of the middle/late Neoglacial interstadial (ca. 1900 B.P.). Although grasses continued to be utilized, a variety of other species, particularly Cheno-Ams, achieved some eminence as major dietary components. The reasons for this subsistence shift are unclear, but may well be related to the introduction of corn and the changes in scheduling, as Hogan (*Appendix X*) has suggested. Cheno-Ams were apparently available throughout the record. Also, it is possible that with the increasing reliance on corn, a more diverse floral inventory was required to compensate for times of loss or poor yield of the corn crop.

[*Acknowledgments*. Special thanks is given to David B. Madsen and Donald R. Currey who offered suggestions regarding interpretation of the record, and Amy Pringle typed the manuscript. Richard E. Fike and Gardiner F. Dalley assisted with the pollen extraction.]

REFERENCES

ANDREWS, J. T., ET AL.

1975   Holocene Environmental Changes in the Alpine Zone, Northern San Juan Mountains, Colorado: Evidence from Bog Stratigraphy and Palynology. *Quaternary Research*, Vol. 5. New York.

ANTEVS, ERNST

1948   Climate Changes and Pre-White Man. *University of Utah Bulletin*, Vol. 38, No. 20, *Biological Series*. Vol. 10, No. 7. Salt Lake City.

1955   Geologic-Climatic Dating in the West. *American Antiquity*, Vol. 20, No. 4, Pt. 1. Salt Lake City.

ASCHMANN, HOMER

1958   Great Basin Climates in Relation to Human Occupance. *University of California Archaeological Survey Report*, No. 42. Berkeley.

BARKLEY, FRED A.

1934   The Statistical Theory of Pollen Analysis. *Ecology*, Vol. 15, No. 3.

BRIGHT, ROBERT C.

1966   Pollen and Seed Stratigraphy of Swan Lake, Southeastern Idaho. *Tebiwa*, Vol. 9. Pocatello.

BRYAN, ALAN L. AND RUTH GRUHN

1964   Problems Relating to the Neothermal Climatic Sequence. *American Antiquity*, Vol. 29, No. 3. Salt Lake City.

CHAMBERLAIN, RALPH V.

1911   The Ethno-Botany of the Gosiute Indians of Utah. *Philadelphia Academy of Natural Sciences Proceedings*, Vol. 63. Philadelphia.

CURREY, DONALD R.

1976a   Late Quaternary Geomorphic History of Pint-Size Shelter, Emery County, Utah. *In* "Pint-Size Shelter," L. W. Lindsay and C. K. Lund. *Antiquities Section Selected Papers*, Vol. 3, No. 9. Salt Lake City.

1976b   Late Quaternary Geomorphic History of Ivie Creek and Sudden Shelter. *In* "Sudden Shelter," J. D. Jennings, A. R. Schroedl, and R. Holmer, *University of Utah Anthropological Papers*. Salt Lake City. (In press)

DENTON, GEORGE H. AND STEPHEN C. PORTER

1970   Neoglaciation. *Scientific American*, Vol. 222, No. 5. New York.

KELLY, ISABEL T.

1964   Southern Paiute Ethnography. *University of Utah Anthropological Papers*, No. 21. Salt Lake City.

LAMARCHE, VALMORE C., JR.

1974   Paleoclimatic Inferences from Long Tree-Ring Records. *Science*, Vol. 183, No. 4129. Washington, D. C.

LINDSAY, LA MAR W.

1976   Pollen Analysis of Sudden Shelter Site Deposits. *In* "Sudden Shelter," J. D. Jennings, A. R. Schroedl, and R. Holmer. *University of Utah Anthropological Papers*. Salt Lake City. (In press)

MADSEN, DAVID B. AND DONALD R. CURREY

1977   Dating Glacial Retreat and Late Quaternary Vegetation Changes, Wasatch Mountains, North Central Utah, U. S. A. Paper presented at the X INQUA Congress, Birmingham, United Kingdom.

MARTIN, PAUL S.

1963   *The Last 10,000 Years: A Fossil Pollen Record of the American Southwest*. University of Arizona. Tucson.

MARTIN, PAUL S. AND FLOYD W. SHARROCK

1964   Pollen Analysis of Prehistoric Human Feces: A New Approach to Ethnobotany. *American Antiquity*, Vol. 30, No. 2. Salt Lake City.

MEHRINGER, PETER J., JR.

1967   Pollen Analysis of the Tule Springs Area, Nevada. *In,* "Pleistocene Studies in Southern Nevada." H. M. Wormington and D. Ellis (eds.). *Nevada State Museum Anthropological Papers*, No. 13. Carson City.

MEHRINGER, PETER J., JR., PAUL S. MARTIN, AND C. VANCE HAYNES, JR.

1967   Murray Springs, A Mid-Postglacial Pollen Record from Southern Arizona. *American Journal of Science*, Vol. 265, No. 9. New Haven.

MEHRINGER, PETER J., JR., AND CLAUDE N. WARREN

1976   Marsh, Dune and Archaeological Chronology, Ash Meadows, Amargosa Desert, Nevada. *In* "Holocene Environmental Change in the Great Basin," R. Elston (ed.). *Nevada Archeological Survey Research Paper*, No. 6. Reno.

PETERSEN, KENNETH L. AND PETER L. MEHRINGER, JR.

1976   Postglacial Timberline Fluctuations, La Plata Mountains, Southwestern Colorado. *Arctic and Alpine Research*, Vol. 8, No. 3. Boulder.

STEWARD, JULIAN H.

1938   Basin-Plateau Aboriginal Sociopolitical Groups. *Smithsonian Institution Bureau of American Ethnology Bulletin* No. 120. Washington, D. C.

WHITING, ALFRED F.

1939   Ethnobotany of the Hopi. *Museum of Northern Arizona Bulletin*, No. 15. Flagstaff.